EXPLORING THE EARTH
UNDER THE SEA

AUSTRALIAN AND NEW ZEALAND ACHIEVEMENTS
IN THE FIRST PHASE OF IODP SCIENTIFIC
OCEAN DRILLING, 2008–2013

EXPLORING THE EARTH
UNDER THE SEA

AUSTRALIAN AND NEW ZEALAND ACHIEVEMENTS
IN THE FIRST PHASE OF IODP SCIENTIFIC
OCEAN DRILLING, 2008–2013

EDITED BY NEVILLE EXON

Australian
National
University

PRESS

ANU PRESS

Published by ANU Press
The Australian National University
Acton ACT 2601, Australia
Email: anupress@anu.edu.au
This title is also available online at press.anu.edu.au

National Library of Australia Cataloguing-in-Publication entry

Title: Exploring the earth under the sea : Australian and
             New Zealand achievements in the first phase
             of IODP Scientific Ocean Drilling,
             2008-2013 / Neville Exon (editor).

ISBN: 9781760461454 (paperback) 9781760461461 (ebook)

Subjects: Scientific Ocean Drilling (IODP)
             Underwater exploration--Australia.
             Underwater exploration--New Zealand.
             Underwater drilling--Australia--Research.
             Underwater drilling--New Zealand--Research.

Other Creators/Contributors:
             Exon, N. F. (Neville Frank) editor.

Cover design and layout by ANU Press

Cover photograph: IODP's great workhorse, the *JOIDES Resolution*, is provided by the United States of America. It works in all the world's oceans and in most water depths, and normally takes continuous cores of sediments or rocks for about 1,000 m below the seabed. Photo courtesy of John Beck of IODP/TAMU.

Contents

Acronyms vii

Preface ix

Foreword xiii

1. The global and regional significance of IODP 1

2. A summary of ANZIC in IODP, 2008–2013 5

3. A brief history of scientific ocean drilling from the Australian
    and New Zealand points of view 11

4. IODP drilling and core storage facilities 19

5. Australian and New Zealand participation in IODP 25

6. Tales from the ship 39

7. A sampler of scientific results 85

8. Education and Outreach 125

9. Major IODP planning workshops 135

10. The future of scientific ocean drilling: International Ocean
    Discovery Program 145

11. Regional proposals drilled since 2013 and to be drilled soon . . .151

12. Broad costs and benefits of Australia's participation in IODP . . .183

13. Major publications by ANZIC science party participants
    arising from 2008 to 2013 expeditions .187

Partners in the first phase of IODP: 2008–2013 213

Acronyms

AIMS	Australian Institute of Marine Science
AIO	ANZIC IODP Office
ANDRILL	Antarctic geological drilling (offshore)
ANSTO	Australian Nuclear Science and Technology Organisation
ANZIC	Australian and New Zealand IODP Consortium
ARC	Australian Research Council
BCR	Bremen Core Repository
CIB	*Chikyu* IODP Board
CORK	subsurface hyrdrogeological observatory
CRISP	Costa Rica Seismogenesis Project
DFAT	Department of Foreign Affairs and Trade
DIICCSTRE	Department of Industry, Innovation, Climate Change, Science, Research and Tertiary Education
DSDP	Deep Sea Drilling Project
DSEWPaC	Department of Sustainability, Environment, Water, Population and Communities
ECORD	European Consortium for Ocean Research Drilling
EPSP	Environmental Protection and Safety Panel
ESO	ECORD Science Operator
GBR	Great Barrier Reef
GCR	Gulf Coast Repository
HUET	Helicopter Underwater Escape Training
IMOS	Integrated Marine Observing Strategy
INVEST	IODP New Ventures in Exploring Scientific Targets

IODP	Integrated Ocean Drilling Program and International Ocean Discovery Program
IODP(1)	Integrated Ocean Drilling Program (2003–2013)
IODP(2)	International Ocean Discovery Program (2013–2023)
JAMSTEC	Japan Agency for Marine-Earth Science and Technology
JCU	James Cook University
JR	*JOIDES Resolution*
JRFB	*JOIDES Resolution* Facilities Board
KCC	Kochi Core Center
LGM	Last Glacial Maximum
LIEF	Linkage Infrastructure, Equipment and Facilities
MARGO	Marine Geoscience Office
mbsf	metres below the sea floor
mbsl	metres below sea level
MEXT	Japanese Ministry of Education, Culture, Sports, Science and Technology
NBC	North Big Chimney
NCAOR	National Centre for Antarctic and Ocean Research
NCRIS	National Collaborative Research Infrastructure Strategy
NSF	US National Science Foundation
OAE	Oceanic Anoxic Events
ODP	Ocean Drilling Program
OM	organic matter
PEAT	Pacific Equatorial Age Transect
SSEs	slow slip events
SST	sea-surface temperatures
TAMU	Texas A&M University
TN	total nitrogen
TOC	total organic carbon
WAIS	West Antarctic Ice Sheet

Preface

Scientific ocean drilling is a very successful large international geoscience program that has gone through a number of iterations since it was founded as the Deep Sea Drilling Project in 1968. In its present configuration, it drills deep holes in all ocean depths, and frequently instruments them, in order to investigate global scientific questions related to Earth's past geological and biological history and the recent situation, as revealed by continuous cores of the sediments and rocks beneath the sea floor.

This publication summarises exciting Australian and New Zealand achievements in the first phase of the Integrated Ocean Drilling Program (IODP(1)) until the end of 2013, when the second phase of the International Ocean Discovery Program (IODP(2)) came into being. The publication carries on the tradition established by *Full Fathom Five; 15 years of Australian Involvement in the Ocean Drilling Program*, which documented Australian involvement in IODP's predecessor the Ocean Drilling Program (ODP) until its end in 2003. *Full Fathom Five* was inspired and led by Elaine Baker and Jock Keene, of ODP's Secretariat in the University of Sydney.

Membership of IODP[1] helps Australia and New Zealand maintain our leadership in Southern Hemisphere marine research. The generosity of our major partners means that we get a marvellous return on our relatively modest investment in the international operational budget, and access to drilling assets worth around US$1.1 billion.

The present volume explains how IODP works and how the Australian and New Zealand IODP Consortium (ANZIC) has functioned within IODP, and outlines our many scientific achievements. Impressions from our scientists from the range of regional expeditions are provided in

---

1   IODP is used as the general term throughout this book to cover both IODP(1) and IODP(2).

Chapter 6, some scientific results are provided in Chapter 7 and a taste of future regional expeditions explored in Chapter 11. Other chapters cover IODP's future, education and outreach, costs and benefits, and a list of major publications involving our scientists. Most of the text was written in 2015 and 2016; later updating has been relatively minor.

Many people have been involved in the production of this book and I am grateful to them all. The ANZIC Governing Council encouraged me from the moment I put the concept of the book to them. The many authors of chapter sections are acknowledged in the titles of those sections – without their enthusiastic support, this book would not have been possible. Michelle Burgess was a skilful and equally enthusiastic editor, who massaged all the contributions, put them into the ANU Press template and provided advice on what might be added and how things might be laid out. Catherine Beasley, our program administrator and my good friend, has been a great help in commenting on ideas and using her considerable word processing skills in helping with the layout where things got tricky. The whole exercise has been a great pleasure to me.

I am very grateful to ANU Press for publishing this book. They selected two international reviewers who provided valuable comments in late 2016: one was Ted Moore of the University of Michigan, Ann Arbor, who has a very long and distinguished history in ocean drilling; the other remained anonymous. They also selected two Australian reviewers: one was Will Howard, of the Australian Chief Scientist's office, who also has a long history in ocean drilling and provided valuable suggestions; the other remained anonymous.

IODP is an exciting program. I have helped steer ANZIC's role in it from 2007 and have thoroughly enjoyed it. It will be hard to step back from this role in late 2017. Two of the many people who played a critical role in the success of the first phase of IODP were the two excellent IODP administrators with whom I worked very closely: Sarah Howgego until the end of 2011 and Catherine Beasley thereafter.

Professor Neville Exon
ANZIC Program Scientist
The Australian National University

The staff of the Australian IODP Office in this period: Neville Exon, Sarah Howgego (until 2011) and Catherine Beasley (from 2012)

Source: Australian IODP Office

The Japanese deep-drilling vessel *Chikyu* at sea. It can core up to 4,000 m below the seabed and in areas where there is a potential danger of striking oil or gas

Source: Photo courtesy of JAMSTEC (Japan Agency for Marine-Earth Science and Technology), which provides the vessel

The European-chartered *Greatship Maya*, which cored the deep-water fossil outer parts of the Great Barrier Reef in 2010 to gather information about the changing character of the reef as the sea warmed and rose 140 m since the coldest period of the last glaciation, about 13,000 years ago

Source: Photo by David Smith (ECORD Science Operator), courtesy of the European Consortium for Ocean Research Drilling (ECORD)

Foreword

The Integrated Ocean Drilling Program (IODP(1)), which followed the highly successful Ocean Drilling Program (ODP), came into existence in 2003 and ended in September 2013, when it was replaced by the International Ocean Discovery Program (IODP(2)). It was designed by a Scientific Planning Working Group, following several large planning meetings with hundreds of participants. The result was *Earth, Oceans and Life: Scientific Investigations of the Earth System Using Multiple Drilling Platforms and new Technologies*, IODP's Initial Science Plan 2003–2013, which was 110 pages long. This outlined a coherent set of scientific investigations of the Earth system, using multiple drilling platforms and new technologies.

Our primary exploration tools are dynamically positioned Japanese and American coring vessels. When the primary vessels are not suitable, the European Consortium for Ocean Research and Drilling (ECORD) charters other coring platforms. The available equipment can take continuous sediment or rock cores in all oceans, at most water depths and up to 5,000 m below the sea floor. IODP has direct access to drilling platforms worth $US1.1 billion and an annual operational budget of about US$180 million. The US, Japan and Europe provide the necessary deep drilling facilities, the core repositories and all the necessary support to help this program function efficiently.

The IODP is a broad and vital collaborative effort. It is the world's largest multinational geoscience program (www.iodp.org). The rationale for this program includes the realisation that the past is often a key to the future of the Earth. Since ocean drilling is the best method of direct sampling below the sea floor, IODP's deep coring provides the means to test global geoscientific theories that are often developed on the basis of remote sensing. New technologies and concepts in geoscience are continuously being developed through IODP. After all, the sediments and rocks (and microbes) beneath the world's oceans contain a remarkable story of how the Earth has worked and how it works now – all of which offer insights into how the Earth may work in the future. These insights can be of great societal relevance in areas such as climate change, natural hazards and mineral deposition. Furthermore, two-thirds of our world is covered by oceans, with some 60 per cent of Australia's and a whopping 95 per cent of New Zealand's territory offshore, in the form of their marine jurisdictions granted under the United Nations Law of the Sea.

In the first phase of IODP(1), which is reported on here, its main research fields were the deep biosphere and ocean floor, including microbes that live in extreme conditions deep beneath the sea floor, and also accumulations of frozen gas hydrates; environmental change processes and effects, including past changes in climate and ocean currents; and solid earth cycles and geodynamics, including continental breakup, large volcanic events, earthquakes and tsunamis.

In 2007, a group of Australian universities and government research agencies applied to the Australian Research Council for funding to join IODP(1) under the Linkage Infrastructure, Equipment and Facilities (LIEF) scheme. Funding was granted in two tranches from 2008 until 2013, when the first phase, the Integrated Ocean Drilling Program ended, although we continued on in the next phase, the International Ocean Discovery Program. New Zealand (never formally members of ODP) joined us in IODP in 2008. Together, we became Associate Members and the Australian and New Zealand IODP Consortium (ANZIC) has never looked back. Fortunately, many of those involved in IODP had worked with the similar ODP, which ended in 2003, so we hit the ground running.

Our achievements are a credit to those who managed to build the case for funding for IODP after an unfortunate hiatus when Australia's ODP membership ended in 2003. The Australian Research Council continued its vital support through the six years 2008–2013, for which we are all very grateful. Of great significance has been the formal entry of New Zealand into scientific ocean drilling for the first time, bringing its breadth of geoscience skills into ANZIC to everyone's benefit. In fact, the ANZIC umbrella has brought together nearly all the major geoscience groups in both countries, and our scientists have taken full advantage of the opportunities provided to carry out research of global significance in partnership with scientists from more than 20 other countries.

It was my pleasure to take over the Chairmanship of the ANZIC Governing Council from Dr Kate Wilson in the middle of 2010, by which time it was running smoothly. Drilling and discovering, IODP is involved in wonderful global science and it was exciting to help steer our involvement.

Dr Geoff Garrett, AO
Chair, ANZIC Governing
Council, 2010–2016

ANZIC Governing Council Chair, Geoff Garrett
Source: Geoff Garrett

1

The global and regional significance of IODP

Neville Exon

Globally, scientific ocean drilling has been a superb international scientific program since its real beginning as the US-funded Deep Sea Drilling Project in 1968, with the first Australian aboard a regional expedition being Gordon Packham from the University of Sydney, on the Southwest Pacific Ocean Leg 21 in 1971–72. Over the years, ocean drilling has moved from exploratory drilling – trying to discover what lies below all the world's oceans – to targeted drilling – seeking to address global scientific problems in suitable locations. It has always taken cores in sediments and rocks deep beneath the ocean floor, starting with the drill ship the *Glomar Challenger*, which took discrete (sequential but intermittent) cores from depths averaging 500 m below the sea floor (mbsf). Now the *JOIDES Resolution* takes continuous cores from depths that are mostly in the range of 500–1,000 mbsf, and the *Chikyu* takes cores that can be as deep as 4,000 mbsf. Very sophisticated continuous wireline logs provide much additional information, and various instruments are left in many holes to record physical and chemical changes over time.

The aims of the present worldwide program are outlined in the International Ocean Discovery Program (IODP(2)) Science Plan for 2013–2023 *Illuminating Earth's Past, Present, and Future*, which was prepared under the aegis of a Science Plan Writing Committee of 14 scientists, including Richard Arculus of The Australian National

University and Peter Barrett of Victoria University of Wellington. It covers many scientific fields under the themes Climate and Ocean Change, Biosphere Frontiers, Earth Connections, and Earth in Motion. Ocean drilling research covers, among other things, the nature of the Earth's mantle and crust, and the nature of the related deep forces that drove or drive the Earth's tectonics; past and future climate change; the history of life as revealed in sedimentary strata; the nature of the extraordinary microbes found deep in the sediments and volcanic rocks beneath the sea floor; and major natural hazards such as earthquakes, tsunamis and submarine landslides. An average two-month IODP expedition recovers thousands of metres of sediments and rocks that provide a wonderful store of highly varied information for subsequent investigation.

Proposals for drilling anywhere in the ocean must address global themes in locations where this can be done particularly well. They are put together by an international group of scientists and are judged by panels and referees simply on their scientific quality and logistical feasibility. We have been involved in excellent regional proposals in recent years, and our region has had a disproportionately high success rate for proposals.

Opportunities exist for researchers (including graduate students) in all specialties – including, but not limited to, sedimentologists, petrologists, structural geologists, palaeontologists, biostratigraphers, palaeomagnetists, petrophysicists, borehole geophysicists, microbiologists and inorganic and organic geochemists. Ocean drilling provides wonderful training for students and researchers at all stages of their careers, and can be a career-changing opportunity. Scientists commonly form lifelong research partnerships with the international scientists they have worked with aboard these vessels, often in fields unrelated to ocean drilling.

The relevance of ocean drilling is not just confined to the oceans. Plate tectonics, which is driven by forces in the Earth's mantle and crust beneath the oceans, has also controlled the past and present nature of the world's land masses and explains, for example, why Australia and New Zealand are so different topographically and in the living and non-living resources that they support. Plate tectonics builds mountain ranges, and explains earthquakes and volcanic areas like those in the Pacific 'Rim of Fire'. Many of the land's rocks were laid down beneath the oceans, and some contain high-value metalliferous ore deposits as well as petroleum formed from marine organisms, so understanding how such resources have formed in the oceans can help exploration for them onshore.

Of course, scientific stratigraphic wells drilled on continental margins provide information of value for petroleum exploration, with many of the world's great petroleum fields offshore.

It should be noted that ocean drilling does not involve just geoscientists. Ocean drilling has shown that very unusual 'extremophile' microbes live deep in the sediments beneath the ocean, and in the warm basalts that have been poured out recently as parts of spreading centres in the world's oceans. Microbiological expeditions make up 10 per cent of ocean drilling expeditions, and are providing much information about these poorly known but globally huge accumulations of extraordinary organisms. Ocean drilling information is also vital to our understanding of past climate change and can constrain the possible scenarios of future climate change. It provides a great deal of information about the drivers of past sea-level rise, and how oceanographic and climatic changes constrained the rates of sea-level change – clearly important information for coastal planners.

The geological hazards of volcanic outbreaks, earthquakes and tsunamis are a major part of ocean drilling research. The recent IODP Expedition 343 off Japan studied the subduction zone and fault that generated the Tohuku-Oki Earthquake of 2011, with its associated devastating tsunami, by not only drilling through the active fault but also instrumenting the drill hole to study changes over time after the earthquake. Such studies and instrumentation lead to better understanding of earthquake hazards, and may well lead to better prediction of future earthquakes. These studies will be continued in the Hikurangi subduction zone east of New Zealand's North Island in 2017 and 2018.

Many hundreds of kilometres of ocean drilling cores and related data are accessible to all scientists from three large core repositories: the Bremen Core Repository at the University of Bremen in Germany (www.marum.de/en/Research/IODP-Bremen-Core-Repository.html), the Gulf Coast Repository located at Texas A&M University (iodp.tamu. edu/curation/gcr/) and the Kochi Core Center at Kochi University in Japan (www.kochi-core.jp/en/).

If policymakers were to ask what our membership of this international ocean drilling program gives us that could not be gained if we were not members and simply made use of ocean drilling research results and access to material, then part of our answer would be that we have helped to drive the science proposals that have brought many expeditions to our region in a way that would be impossible for non-members and that, without our ideas and hard work within the system, many of these expeditions would not have occurred or been scheduled. These activities have provided or will provide insights into questions of national and regional significance, which could not have been obtained in any other way. Among those being addressed are the geological history of the Great Barrier Reef, the Australian Monsoon and the Antarctic margin claimed by our countries. In addition, the study of the earthquake- and tsunami-prone Hikurangi subduction zone east of New Zealand's North Island is of great societal relevance.

Another answer to such a policy question is that having our students and scientists involved with the expeditions and their aftermath is wonderful training and allows the interchange of information at the highest levels of international science. Furthermore, these people form international alliances to address other aspects of geoscience or microbiology, to the benefit of both Australia and New Zealand.

2

A summary of ANZIC in IODP, 2008–2013

Neville Exon

IODP is the world's largest international scientific geoscience program, with a yearly operational budget of about US$180 million and 26 participating nations at the end of 2013. IODP deploys two large drilling vessels and other drilling platforms on scientific expeditions. Australian and New Zealand scientists, through ANZIC, have at least one position on each expedition. The coring platforms and core repositories are provided by foreign entities, and we obtain access to them by paying membership contributions. The platforms are in three categories: the workhorse vessel *JOIDES Resolution*, the deep-drilling vessel *Chikyu* and the alternative platforms provided by the European Consortium for Ocean Research Drilling (ECORD).

In 2013, 14 Australian universities and four government research agencies were part of the ANZIC consortium, along with two government agencies and two universities from New Zealand. Our annual funding of about AU$2.5–3 million came from the Australian Research Council and our Australian and New Zealand members, and US$1.5–1.8 million of this went toward our IODP memberships.

Through the ANZIC consortium, Australia and New Zealand are important players in the exciting IODP project, with the eager participation of many scientists. With our partners including nearly all the major science organisations on Earth (those from Brazil and Israel being the latest to join), we have direct access to, and active engagement with, the world's best scientific minds. We estimate that from early 2008 until late 2013, at least 80 Australians and 20 New Zealanders have worked on ocean drilling science, and our scientists' voices are respected in key international decision-making committees. Our member universities and research organisations have worked exceedingly well together, and with their international counterparts. Our contribution to scientific ocean drilling has been substantial, with the involvement of Australians in 9 per cent and New Zealanders in 3 per cent of all refereed publications at the end of 2013 – all for a 1 per cent contribution to the overall US$180 million annual international operational budget.

Membership of IODP is critical to helping us maintain our leadership in Southern Hemisphere marine research. The Australasian region saw five IODP expeditions in 2009 and 2010 (with an Australian co-chief scientist on one of them), another four occurred in 2015 and 2016, and 10 have been scheduled for 2017 and 2018. We have at least one scientist on all IODP expeditions and, on regional expeditions to which our scientists have provided much input, that number can increase to four or five, often including a co-chief scientist.

The generosity of our major partners means that we get a marvellous return on our very modest investment in the international operational budget, and access to assets worth around US$1 billion. Note that an average two-month IODP expedition recovers thousands of metres of sediments and rocks that provide a wonderful store of highly varied information for subsequent investigation.

From 2008 to 2013, ANZIC put 34 scientists on 25 IODP expeditions (of a possible 32), and another two were part of a science party but did not go to sea. Of the 35 scientists directly involved, 29 were based in Australia and seven in New Zealand. We also put a New Zealand science communicator aboard one expedition.

The expeditions were split geographically as follows:

- Western Pacific Ocean: 14 (7 in our region; 7 *Chikyu* in northwest Pacific)
- Eastern Pacific Ocean: 6
- Southern Ocean: 1 (in our region)
- Atlantic Ocean: 3

The expeditions could be split roughly as follows:

- Deep Biosphere and Subseafloor Ocean: 2 (1 in our region)
- Environmental Change, Processes and Effects: 9 (4 in our region)
- Solid Earth Cycles and Geodynamics: 14 (1 in our region; 5 *Chikyu* in northwest Pacific)

The expeditions of most immediate scientific interest to us were those in our region:

- Canterbury Basin Sea Level Expedition 317 was drilled by the *JOIDES Resolution* east of New Zealand's South Island, and included the deepest sedimentary hole (2,000 m below the sea floor (mbsf)) ever drilled by the ship without re-entry to the drill hole. It has provided excellent information about global sea-level rise and fall in the last 15 million years.
- Wilkes Land Glacial History Expedition 318 was drilled by the *JOIDES Resolution* north of the Australian Antarctic Territory, and revealed a great deal about the change from a warm Antarctica 33.5 million years ago to its present state, including the dramatic cooling of the ocean. For example, it showed that palm trees grew on the Antarctic margin 55 million years ago.
- Great Barrier Reef Environmental Changes Expedition 325 was drilled by the *Greatship Maya* into ancient reef platforms seaward of the Great Barrier Reef. It has provided material for a unique study of the ancestral reefs at the peak of the last glaciation (then 140 m below the present sea level) and those that formed as sea level rose as the Earth warmed thereafter. It has provided important evidence that sea level rose at highly variable rates.
- The South Pacific Oceanic Gyre Subseafloor Life Expedition 329 was drilled by *JOIDES Resolution* in the south-central Pacific. It was designed to test the depth to which microbes survive and examine their nature in one of the lowest productivity regions on Earth. It showed that microbes were relatively rare here but did exist to considerable depths. The concentration profiles of oxygen and nitrate demonstrate that the rate of microbial respiration is generally extremely low.
- The Louisville Seamount Trail Geodynamics Expedition 330 was drilled by *JOIDES Resolution* and covered a northwest-trending seamount trail east of the Tonga Trench. It was designed to test whether the Louisville hotspot had remained fixed over time, and whether

the nature of volcanic seamounts changes over time. It showed that the hotspot had remained fixed and that there was petrologically homogeneous volcanism through time.

As regards publications, membership strongly affects output, but there is also a time lag of some years. Neither Australia nor New Zealand was a member in the early days of ocean drilling, although Australia was a member of ODP from 1989 to 2003, and Australia and New Zealand joined IODP only in 2008. The number of actual publications from IODP expeditions since then is only a small proportion of what we expect in the longer term. It is worth noting that 11.4 per cent of ocean drilling publications had ANZIC authors from 2003, when IODP started, to 2013. The total number of recorded refereed DSDP–ODP–IODP publications with Australian and/or New Zealand authors up to 2013 is 3,451, or 12 per cent of all publications.

Two major international planning workshops were held during this period – the IODP New Ventures in Exploring Scientific Targets (INVEST) workshop in Bremen, Germany, in 2009, and the Chikyu+10 workshop in Tokyo in 2013 – and ANZIC scientists were heavily involved in both. Two major regional Australian-inspired workshops were also important. The first such workshop was the Indian Ocean IODP Workshop held in Goa, India, in 2011. This workshop brought scientists interested in ocean drilling in this region together to plan IODP proposals that have led to a series of expeditions in the Indian Ocean in 2015 and 2016. The second was the Southwest Pacific Ocean IODP Workshop held in Sydney in 2012. It also led to a number of proposals that have or will come to fruition with Pacific Ocean drilling in 2016, 2017 and 2018.

There have been port calls with associated ship visits and publicity in Townsville, Hobart, Wellington and Auckland by the *JOIDES Resolution,* and in Townsville by the *Greatship Maya*. The Hobart port call was attended by local scientific leaders and the then Minister for Innovation, Industry, Science and Research, the Honourable Kim Carr, and many others. The New Zealand port calls featured extensive visits and excellent media publicity.

Outreach activities remain central to ANZIC's mission; in both 2012 and 2013, we funded 20 university undergraduate students to attend an ANZIC Marine Geoscience Masterclass in Perth, with the aim of inspiring the next generation of scientists to work in this exciting area of research. Their feedback was very positive.

The Allen Consulting Group in Canberra carried out a review of Australia's participation in the Integrated Ocean Drilling Program that was finalised in March 2013 (iodp.org.au/publications/independent-review-of-australian-participation-in-integrated-ocean-drilling-program/). They concluded:

> that the benefits to Australia of direct membership of the IODP consortium far exceed the modest costs of participation. Moreover it would be detrimental to Australia's interests not to be a member of the next phase of scientific ocean drilling. Participation in this next phase is well aligned with current government policy as articulated in the 2012 National Science Investment Plan, the aspirations of the Australia in the Asian Century White Paper and Australia's policy of fostering international scientific collaborations (see Chapter 12 for more about this review).

The first 10-year phase of IODP, IODP(1), ended in September 2013, with another 10-year phase of ocean drilling from late 2013 to 2023 under the new name International Ocean Discovery Program (IODP(2)) with the same acronym, IODP. The structure of the new program is much looser than the previous one, with those who provide the vessels – the US, Japan and Europe – having ultimate control of the associated programs. Australian institutions received funding for two years (2014 and 2015) under an ARC/LIEF grant and this was later renewed for 2016 to 2020, and New Zealand remains in ANZIC for that period.

3

A brief history of scientific ocean drilling from the Australian and New Zealand points of view

Neville Exon

A full review of Australia's involvement in ocean drilling since its onset in 1968 through the Deep Sea Drilling Project (DSDP) to the 1985–2003 Ocean Drilling Program (ODP), plus the initial part of the 2003–2013 Integrated Ocean Drilling Program (IODP) was provided by Neville Exon's 2010 paper: Scientific drilling beneath the oceans solves earthly problems, published in the *Australian Journal of Maritime and Ocean Affairs* 2(2), 37–47. Much of this short summary is drawn from that paper.

A broader, briefer review was provided by Deborah K. Smith, Neville Exon, Fernando J.A.S. Barriga and Yoshiyuki Tatsumi, 2010: Forty years of successful international collaboration in scientific ocean drilling, published in *Eos*: *Transactions American Geophysical Union* 91(43), 393–404 (doi.org/10.1029/2010EO430001).

Early days: The Deep Sea Drilling Project

Between 1968 and 1983, the DSDP began the study of the deep ocean's sediments and rocks using the *Glomar Challenger*. DSDP was funded by the US National Science Foundation but welcomed foreign scientists, including Australians and New Zealanders, to its drilling campaigns.

This ship started to build a story about what was happening and had happened in the 70 per cent of the Earth's crust that lies beneath the oceans. Its major achievements included:

- Drilling and dating the oceanic basalts that form on the sea floor as continents drift apart and are later covered in sediment, thus contributing greatly to the concept of plate tectonics.
- Proving that the oceanic rocks existing at present have all formed in the last 200 million years, and showing that such rocks are continuously poured out at mid-oceanic ridges and destroyed at oceanic trenches. Continental rocks, by contrast, can be billions of years old. Many of these 'continental rocks' are, in fact, ancient sedimentary and volcanic rocks that formed in the ocean but have been accreted to the continents.
- Providing a detailed history of the climate and oceanographic changes that have affected the world's oceans in the last 200 million years.

Figure 3.1. *Glomar Challenger* at sea
Source: *JOIDES Resolution* Science Operator, Texas A&M University

This was the first-generation exploration phase of ocean drilling, with holes being drilled in most parts of the world's oceans to test existing ideas and also to see what was actually there. All holes were spot-cored rather than continuously cored.

DSDP drilled eight legs in the Australasian region in two phases. In the first phase, the first regional expedition was Leg 21 in 1973, with Gordon Packham of the University of Sydney aboard, and the last was Leg 33 in 1976. Both these legs happened to be in the Southwest Pacific, but others were elsewhere in the region. This exploratory phase set the scene for the problem-solving expeditions that came later, including the final DSDP expedition in our region in 1986, which was the oceanographic Leg 90 in the Southwest Pacific under the International Phase of Ocean Drilling (IPOD), a predecessor to ODP.

A great deal was learned about the volcanic ridges and the intervening sedimentary basins that characterise the sea floor in our region, and about its plate tectonic history. For example, it was shown that other Gondwanan continents broke away from Australia, starting about 160 million years ago, with Australia moving north from Antarctica in the last 90 million years.

Early maturity: The Ocean Drilling Program

In 1985, a larger and more capable drilling vessel, the *JOIDES Resolution*, replaced the *Glomar Challenger* in the new Ocean Drilling Program (ODP), and continuous coring becoming normal. This was a phase of ocean drilling that aimed to solve global scientific problems, unlike the DSDP's curiosity-driven exploration. Many scientific questions of global significance were addressed, and the understanding of our geological framework increased greatly.

This phase of ocean drilling was still largely funded by the US, but considerable funds were also provided by other countries, especially European countries and Japan. Although scientists from member countries took up most positions on vessels, scientists from countries that were not members, especially where drilling was taking place, also participated. Australia joined ODP in 1988 in a consortium with Canada, Korea and Taiwan, and Australian scientists were heavily involved both before Australia joined and afterwards. New Zealand never joined ODP, but its scientists were often involved; for example, on the oceanographic Leg 181 to the Campbell Plateau and Chatham Rise region.

Figure 3.2. *JOIDES Resolution* travelling through the Panama Canal in ODP days

Source: *JOIDES Resolution* Science Operator, Texas A&M University

There were 17 two-month ODP expeditions in the Australasian region, starting with Leg 119 in Prydz Bay, south of the Kerguelen Plateau, in 1988, and ending with Leg 194 on the Marion Plateau off eastern Queensland in 2001. Australian and New Zealand scientists were heavily involved in writing proposals, and seven Australians became co-chief scientists on expeditions. By the time ODP ended in 2003, 71 Australian scientists had participated in expeditions, and seven Australian scientists had acted in the key position of co-chief scientist. At the end of ODP, Elaine Baker and Jock Kenne of the Australian ODP Office in Sydney published an excellent review of Australian achievements in ODP entitled *Full Fathom Five: 15 Years of Australian Involvement in the Ocean Drilling Program*.

The Australian geoscience research vessel *Rig Seismic* was instrumental in conducting some of the detailed site surveys that were essential for successful ODP proposals. When *Rig Seismic* was disposed of in 1998, Australia unfortunately no longer had the world-class seismic profiling capability needed for most site survey work. Fortunately, a new world-class Australian Marine National Facility, the *Investigator*, came into operation in 2014 and it has been gathering data for future IODP proposals.

Maturity: The Integrated Ocean Drilling Program

From 2003 to the present day, we have had two phases of IODP, Integrated Ocean Drilling Program (IODP(1)) and International Ocean Discovery Program (IODP(2)), with a major rebuild and refit of *JOIDES Resolution* early in the cycle, which greatly increased its laboratory capabilities and improved its living quarters, among other things. In addition, the deep drilling capability of the Japanese drill ship *Chikyu* and specialist platforms provided by the Europeans have greatly widened the scope of research activities. The European-chartered drilling vessels were invaluable for occasional expeditions when neither of the other two vessels was suitable. This meant that drilling in the Arctic Ocean and in the Great Barrier Reef, for example, was now possible.

When ODP ended there was no easy mechanism for Australia to join IODP, but Helen Bostock of ANU was commissioned and funded (from money carried forward from Australia's ODP consortium) to find a way forward. When a major IODP workshop was held in Hobart, with international and national participants, there was clearly great interest from Australian research institutions, and our efforts redoubled after that. In January 2005, a proposal for Australia to join IODP, as part of an Asian consortium with South Korea and potentially India and Taiwan, was put forward to the National Collaborative Research Infrastructure Strategy (NCRIS), under the guideline to 'provide access to infrastructure'. After some changes during the assessment process, the IODP proposal was included in the Integrated Marine Observing Strategy (IMOS) capability in February 2006. Unfortunately, it was distinctly different from other proposals within the IMOS capability and was excluded from final IMOS funding.

Figure 3.3. *JOIDES Resolution* leaving Honolulu in 2008, soon after its refit
Source: *JOIDES Resolution* Science Operator, Texas A&M University

Later in 2006, an Australian IODP steering group decided that we should work toward an IODP Australian Research Council (ARC) bid, and Neville Exon was commissioned and funded to steer a bid to be submitted in early 2007, for funding to commence in January 2008. Several meetings of scientists representing a number of Australian institutions were held, and a Canberra-based geoscience group undertook to lead this project. There were many important players, but Richard Arculus, Patrick De Deckker and Neville Exon, all of ANU, formed a core group. The group convinced the ANU Deputy Vice Chancellor of Research, Professor Lawrence Cram, to provide $100,000 per year from central funds if the bid succeeded. Things then moved fast, with 18 Australian institutions joining the ARC bid, and New Zealand agreeing to join a consortium if we succeeded. Richard Arculus became the lead chief investigator on the bid with another 20 chief investigators from universities or principal investigators from government research agencies also signing.

In late 2007, we were informed that the ARC/LIEF bid had succeeded and, with the funds from ARC and the Australian partners, Australia was able to join IODP as an Associate Member. New Zealand then joined to make the Australian and New Zealand IODP Consortium (ANZIC). Australia and New Zealand joined IODP in 2008 and have been very active members since. ANZIC has had three co-chief scientists thus far, and will have had another six by the time *JOIDES Resolution* leaves our region in 2018. Altogether, 35 ANZIC scientists and one science communicator have been part of various expedition science parties up to the end of 2013, and many more since. Details of our involvement are set out in Chapter 5.

Thus far there have been three phases of IODP drilling in the Australasian region. The initial phase consisted of five expeditions that started with Expedition 317 in the Canterbury Basin, east of New Zealand, in late 2009, and ended with Expedition 330 to the Louisville Seamount Trail, northeast of New Zealand, in early 2011. The next phase of 'regional' drilling, 10 expeditions in the Indian Ocean, was triggered by the Indian Ocean IODP Workshop in Goa in 2011. There were five expeditions in the area between India and Australia, starting with Expedition 353 studying the Indian Monsoon in late 2014, and ending with Expedition 362 studying the Sumatra seismogenic zone in late 2016. This was immediately followed by the Western Pacific Warm Pool Expedition 363, which was stimulated by the Indian Ocean Workshop.

In 2012, the Southwest Pacific IODP workshop was held in Sydney, and it too stimulated a whole series of proposals, nearly all with New Zealand leadership, with six expeditions to be drilled in 2017 and 2018 in the third phase of regional drilling. These will start with Expedition 371 to the Lord Howe Rise in mid-2017 and end with Expedition 376 to the Brothers Volcano in mid-2018.

The Australasian IODP Regional Workshop, held in Sydney in June 2017, is expected to help initiate another round of IODP drilling in this region in the early 2020s.

4

IODP drilling and core storage facilities

Neville Exon

As the knowledge obtainable from ocean drilling is various and extensive, its end-users are similarly various and extensive. Scientific ocean drilling in its various forms has drilled thousands of continuously cored drill holes in all the world's oceans, and all the cores are available to scientists everywhere. It has helped prove the theory of plate tectonics and it is the key information source for past changes in global oceanography and climate. It is also a major information source for the processes that control oceanic volcanism and seabed mineralisation. It addresses the formation of continental margins, oceanic plateaus and island arcs. Ocean drilling works uniquely in investigating subduction zones that generate earthquakes and tsunamis. For example, after the enormous 2011 Japanese earthquake and tsunami, the IODP vessel *Chikyu* drilled through and instrumented the generative fault, and was able to show that nearly all stress had dissipated and that another such earthquake from that fault was unlikely for many years.

Many ocean drilling expeditions have drilled deep stratigraphic core holes in sedimentary basins on continental margins for primarily scientific reasons, but the results are widely used by petroleum exploration companies and government agencies interested in the petroleum potential of these basins. The ages and nature of the sediments, the wide set of physical and chemical data, and the industry-style wireline logs provided by IODP are part of a petroleum explorer's tools of trade.

IODP drilling facilities

IODP uses various drilling platforms to access different subseafloor environments during research expeditions (see Figure 4.1 below). Three science operators in the US, Japan and Europe manage these platforms. The main aim of these platforms is to take continuous sediment or rock cores down to varying depths below the sea floor, plus geophysical measurements *in situ* and in the recovered cores. Ancillary information comes from installing strings of observatory equipment at varying depths below the sea floor to measure, for example, the physical characteristics of the sediments or rocks and the stresses working on them, seismic activity, and the chemical composition of the pore water. A huge amount of information can be obtained from the recovered sediments, rocks and microbiological samples, and from the observatory equipment.

The US provides the *JOIDES Resolution* as IODP's workhorse, and it uses riser-less drilling technology with seawater as the primary drilling fluid, which is pumped down through the drill pipe. The seawater cleans and cools the drill bit and lifts cuttings out of the hole, leaving them in a pile around the hole. The vessel can drill in water depths of 70 to 6,000 m, and generally drills holes to less than 1,000 m below the sea floor. If it is simply drilling sediments and relatively soft sedimentary rock, it can recover 5,000 m or more of core during a two-month expedition. If it is drilling hard igneous or metamorphic rocks then that figure is greatly reduced.

Japan provides the larger and much more complex *Chikyu* to drill deeper than the *JOIDES Resolution*, often in areas where over-pressured oil or gas might be encountered. It has a marine riser system to maintain the pressure balance within the borehole, which includes an outer casing that surrounds the drill pipe to provide return-circulation of dense drilling fluid. A blowout preventer on the sea floor protects the drill works from uncontrolled pressure release. This technology is necessary for drilling several thousands of metres into the Earth.

Europe provides alternative platforms for areas that are not suitable for the other two vessels. Such areas include the Arctic and Antarctic where icebergs and floating ice require specialist vessels, and shallow water regions (e.g. among coral reefs). Such drilling has used a fleet of icebreakers including an ice-breaking drill ship, or a relatively small commercial vessel with a wireline rig, or a jackup rig, or a conventional oceanographic vessel and a seabed drill, depending on the task at hand.

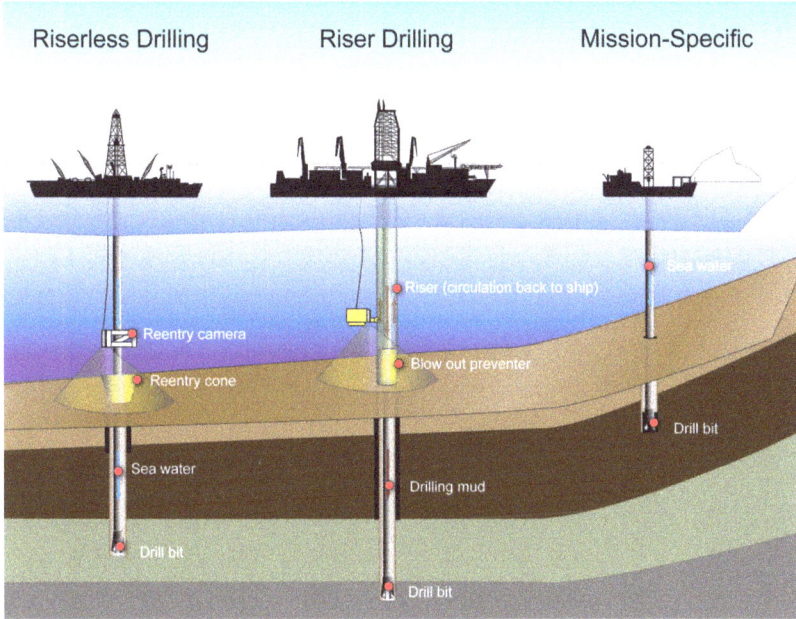

Figure 4.1. IODP vessels and their varied capabilities

JOIDES Resolution carries out standard riserless drilling, *Chikyu* can drill much deeper and uses a marine riser to help control rising fluids, and the Europeans provide a variety of vessels for non-standard activities.

Source: US Science Support Program

All ocean drilling cores (DSDP, ODP and IODP) are kept in cool conditions (4°C) in three repositories: one in the US, one in Germany, and one in Japan (www.iodp.org/resources/core-repositories). Most cores from our region are stored in Japan. A total of about 400 km of drill core, the result of 45 years of ocean drilling, are stored in these repositories, along with biological samples stored in a freezer at –20°C or in liquid nitrogen (–196°C). Members of the science party for each expedition have sole access to material for a one-year moratorium period after they obtain their samples. Once the moratorium for each expedition is over, access to core and other relevant material is provided to any bona fide scientist on the basis of a high-quality research proposal and the agreement to publish the results, assuming the proposed work does not directly clash with science party activities. People can either visit the repository to examine and/or select material, or order material on the basis of online reports and images. A curatorial advisory board makes final decisions regarding distribution of all samples.

The core repositories are:

- the Bremen Core Repository (BCR) located at MARUM at the University of Bremen, Germany
- the Gulf Coast Repository (GCR) located at Texas A&M University in College Station, Texas
- the Kochi Core Center (KCC) located at Kochi University in Kochi, Japan.

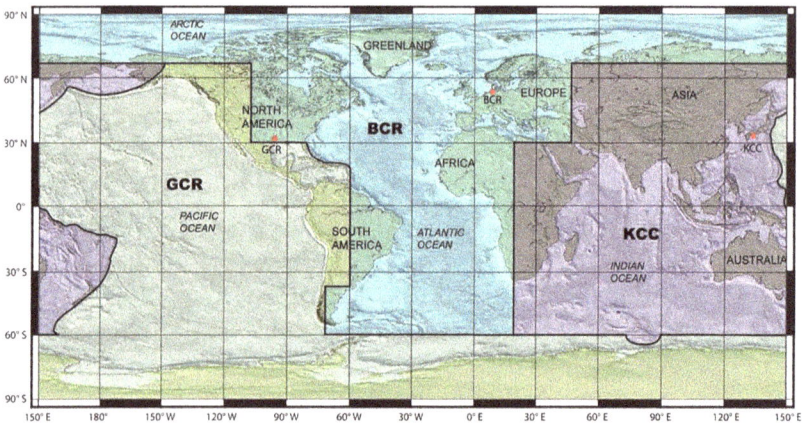

Figure 4.2. Map showing areas of responsibility for IODP core repositories

Cores are distributed largely by geographic area as shown in this map: BCR = Bremen; GCR = Gulf Coast; KCC = Kochi.

Source: Ursula Röhl, adapted from Firth, J.V., Gupta, L.P. and Röhl, U. (2009) New focus on the Tales of the Earth – Legacy Cores Redistribution Project Completed. *Scientific Drilling* 7. 31–33. doi.org/10.2204/iodp.sd.7.03.2009. [Map: 15 March, 2016]. Retrieved from marum.de/en/Cores_at_BCR.html

The BCR stores cores from the Atlantic Ocean, the Mediterranean and Black seas, and the Arctic Ocean. In 2015, the repository contained about 154 km of deep-sea cores from 87 expeditions, in about 250,000 boxes, which are sampled and analysed by national and international working groups. Around 200 scientists visit the repository annually, sometimes working on the cores in weeks-long sampling meetings. As many as 50,000 samples per year are taken by guests and by the repository staff. The repository is an important contact point for scientists from all over the world (there were more than 4,000 visitors by early 2015) and therefore significantly contributes to the exchange and transfer of marine science knowledge, leading to many international cooperations and scientific interactions.

Figure 4.3. Part of the Bremen Core Repository illustrating the enormous size of these facilities

Source: MARUM, University of Bremen, Germany

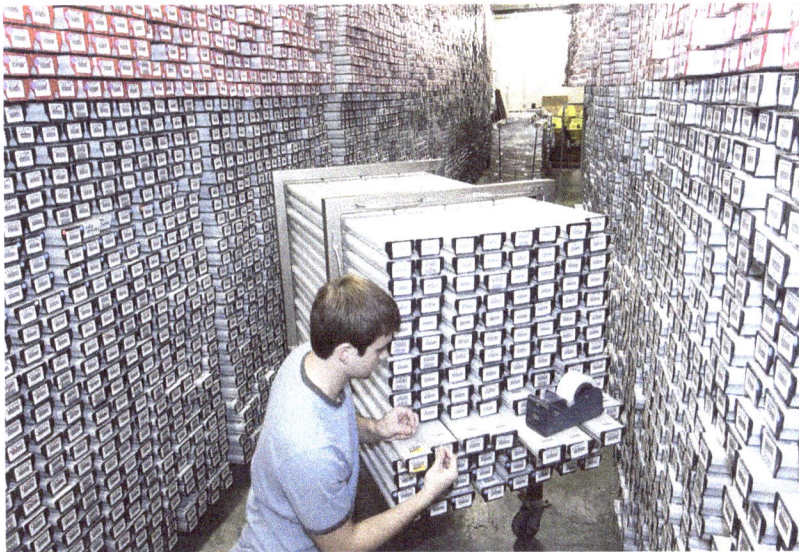

Figure 4.4. Young geologist sampling at Gulf Coast Repository

Source: Courtesy of *JOIDES Resolution* Science Operator

The GCR is located on the Texas A&M University campus in College Station, Texas. This repository stores cores from the Pacific Ocean, the Caribbean Sea and Gulf of Mexico, and the Southern Ocean. A satellite repository at Rutgers University houses New Jersey/Delaware land cores 150X and 174AX. By early 2015, the GCR housed over 132 km of core in approximately 15,000 square feet of refrigerated space. In addition, the GCR stores thin section, smear slide, and residue collections, all of which are available to scientists for study.

The KCC stores cores from the Indian Ocean and the western Pacific Ocean. Cores are stored in three large reefers. These reefers are fitted with mobile core racks in order to enhance their storage capacity. Each core rack has mesh-like structures to facilitate storage of core sections in D-tubes. In early 2015, the core racks allowed storage of about 155,000 D-tubes, which provides sufficient space for storage of about 117 km of core. There is a large freezer to store biological samples at –20°C and a special facility to store samples in liquid nitrogen. Besides these, there are air-conditioned reefer containers that are used for storage of salt cores, core catcher samples and residues at room temperature and low humidity conditions. The facility was expanded in 2015.

Figure 4.5. Kochi Core Center
Source: Japan Agency for Marine-Earth Science and Technology (JAMSTEC)

5

Australian and New Zealand participation in IODP

Neville Exon

Australia and New Zealand have been partners in the ANZIC consortium within IODP since the consortium's establishment in 2008 (www.iodp.org.au; www.gns.cri.nz/Geodiscovery/In-the-Ocean/IODP-National). Our scientists (geoscientists and microbiologists) are making important scientific contributions, and coring expeditions in our region and elsewhere have improved and will continue to improve our understanding of global scientific questions. IODP brings our scientists into contact with research teams from around the world, and post-cruise research activities often extend far beyond IODP ventures.

Membership of IODP has helped us maintain our leadership in Southern Hemisphere marine research. Our geography, climate, oceanography and plate tectonics make our region vital when addressing various global science problems. The Australasian region has seen a great deal of ocean drilling since 1968, including five IODP(1) expeditions from 2009 to 2013; more IODP(2) expeditions have occurred and are about to occur in the region from 2015 to 2018. ANZIC scientists gain through shipboard and post-cruise participation, by building partnerships with overseas scientists, by being research proponents and co-chief scientists who can steer programs and scientific emphasis, and by early access to key samples and data.

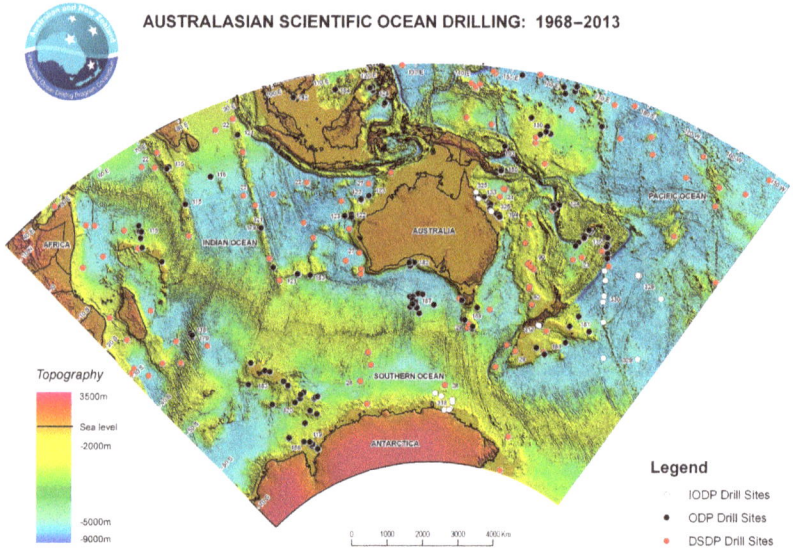

Figure 5.1. Map of Australasian Scientific Ocean Drilling: 1968–2013
Source: Courtesy of Geoscience Australia

Our postdoctoral and doctoral students have an opportunity to train in areas of geoscience and microbiology that cannot be studied in any other way. A new initiative was a Marine Geoscience Masterclass for advanced undergraduate students from all our university partners, held in Perth in late 2013; four more were held by the end of 2016.

We successfully worked toward Australian and New Zealand continuation in the new IODP(2) phase beginning in October 2013, and that was later funded in October 2015 through to 2020. In both countries, this involved using the successes of IODP, and the exciting science outlined in the new Science Plan, to convince scientific and funding institutions that ongoing support is essential.

ANZIC membership

Australia has been the major financial contributor to ANZIC, with benefits being shared between Australia and New Zealand. Australia paid an annual membership fee of US$1.4 million throughout IODP(1). The members of ANZIC have varied slightly over time, as shown in Table 5.1. Initially, New Zealand was a full national member in the ANZIC consortium, paying its share of ANZIC's IODP Associate

Membership directly to IDOP, but government funding was reduced in 2011, and thereafter New Zealand members paid to join the Australian consortium within ANZIC.

In 2013, the Australian IODP Consortium consisted of 14 universities, four government agencies and one marine geoscience peak body. The New Zealand membership consisted of GNS Science and three universities. Our strength is in our breadth of support in the geoscience community.

Table 5.1. Australian and New Zealand ANZIC members: 2008–2013

Australian Institutions	City	Years of Membership
Australian Institute of Marine Science (AIMS)	Townsville	2008–2013
The Australian National University (ANU)	Canberra	2008–2013
Australian Nuclear Science and Technology Organisation (ANSTO)	Sydney	2008–2013
CSIRO Earth Science and Resource Engineering	Perth	2008–2013
Curtin University	Perth	2009–2013
Geoscience Australia	Canberra	2013
James Cook University	Townsville	2008–2013
Macquarie University	Sydney	2008–2013
Monash University	Melbourne	2008–2013
Queensland University of Technology	Brisbane	2009–2013
University of Adelaide	Adelaide	2008–2012
University of Melbourne	Melbourne	2008–2013
University of Newcastle	Newcastle	2008–2011
University of New England	Armidale	2012–2013
University of Queensland	Brisbane	2008–2013
University of Sydney	Sydney	2008–2013
University of Tasmania	Hobart	2008–2013
University of Technology Sydney	Sydney	2011–2013
University of Western Australia	Perth	2008–2013
University of Wollongong	Wollongong	2008–2013
Marine Geoscience Office (MARGO)	Canberra	2008–2013
New Zealand Institutions	City	Years of Membership
GNS Science	Wellington	2008–2013
University of Auckland	Auckland	2012–2013
University of Otago	Dunedin	2008–2013
Victoria University of Wellington	Wellington	2008–2013

The chief and principal investigators in the ARC/LIEF grant varied somewhat over time, as shown in Table 5.2.

Table 5.2. ANZIC chief and principal investigators: 2008–2013

Investigator	Institution	Period
Richard Arculus	ANU	2008–2013
Mark Barley	University of Western Australia	2008–2013
Jochen Brocks	ANU	2008–2013
Ben Clennell	CSIRO	2008–2013
Alan Cooper	University of Adelaide	2008–2012
Patrick De Deckker	ANU	2008–2013
John Dodson	ANSTO	2008–2013
Russell Drysdale	University of Newcastle/Melbourne	2008–2013
Neville Exon	ANU	2008–2013
Chris Fergusson	University of Wollongong	2008–2013
Janet Hergt	University of Melbourne	2008–2013
Will Howard	University of Tasmania/ANU	2008–2013
Peter Kershaw	Monash University	2008–2013
Campbell McCuaig	University of Western Australia	2008–2013
Dietmar Müller	University of Sydney	2008–2013
Suzanne O'Reilly	Macquarie University	2008–2013
Ian Poiner	AIMS	2008–2013
Paolo Vasconcelos	University of Queensland	2008–2013
Jody Webster	JCU/University of Sydney	2008–2013
Chris Yeats	CSIRO	2008–2013
Lindsay Collins	Curtin University of Technology	2009–2013
Gary Huftile	Queensland University of Technology	2009–2013
Greg Skilbeck	University of Technology Sydney	2011–2013
Mike Coffin	University of Tasmania	2012–2013
James Daniell	James Cook University	2012–2013
Chris Hollis	GNS Science, Wellington	2012–2013
Tim Naish	Victoria University of Wellington	2012–2013
Gary Wilson	University of Otago, Dunedin	2012–2013
Alan Baxter	University of New England	2012–2013
Ingo Pecher	University of Auckland	2013
Andrew Heap	Geoscience Australia	2013

ANZIC Governing Council

ANZIC is controlled by its governing council, a steering committee that directs broad policy. This council provides scientific and financial oversight of Australian activities, including those of the ANZIC IODP Office (AIO) and the Science Committee, in conjunction with the ANU Delegate (Director, Research School of Earth Sciences), as ANU handles the consortium funds and hosts the AIO. Council membership favours the organisations that contributed most in terms of funding; they have permanent places on council. Many of the members were rotated though the council over time (Table 5.3). The ANU Delegate was Professor Brian Kennett in the early years, Professor Andrews Roberts from 2010 to 2012, and Professor Ian Jackson in 2013. All took an active and supportive interest in IODP, and two were members of governing council.

Two excellent chairs have controlled the ANZIC Governing Council. The first chair, Dr Kate Wilson (then Director of CSIRO's Wealth from Oceans Flagship), is a molecular biologist by training and a marine scientist in fact. She helped establish a firm organisational structure for ANZIC, enabling us to run smoothly remarkably quickly. She continued in this role until 2010. Dr Geoff Garrett (former Chief Executive of CSIRO, then Chief Scientist of the Queensland Government) succeeded Dr Wilson and continued to steer ANZIC down what was by then a reasonably well-defined path. New Zealander Dr Chris Hollis of GNS Science in Wellington was a vital part of the ANZIC leadership throughout. By the end of 2013, meetings of the governing council had been held in all cities with IODP members, except Adelaide and Newcastle.

Figure 5.2. Dr Kate Wilson, ANZIC Chair until mid-2010
Source: Kate Wilson

The major crisis that faced ANZIC under Dr Kate Wilson's leadership was the 2009 collapse in the Australian exchange rate against the American dollar, which was exceedingly serious given that a large part of our budget went to pay our membership in American dollars (at one stage AU$1 was worth 55 US cents). Fortunately, both ARC and our members stepped in with additional funds to cover the shortfall. Because the Australian and New Zealand dollars rallied soon afterward, we had a substantial budget surplus by 2012, and greatly reduced our membership contributions in 2013.

Dr Geoff Garrett took over from Kate as ANZIC Chair in mid-2010. He had been an important member of a major US Triennial Review of the IODP program earlier in that year, so came to the role very well informed. This review helped revise the structure and planning of the first phase of IODP to improve efficiency. In the years from 2010, Geoff worked to more widely publicise IODP and the benefits for ANZIC scientists of being involved in this great global scientific program. Geoff stepped down as Chair of the governing council at the end of 2016 and was replaced by the eminent marine scientist Ian Poiner. We owe Geoff a huge debt of gratitude.

Figure 5.3. Dr Geoff Garrett, ANZIC Chair from mid-2010

Source: Geoff Garrett

Table 5.3. ANZIC Governing Council: 2008–2013

Member	Position	Institution	Expertise
In 2013			
Geoff Garrett from 2010	Chair	Chief Scientist, Queensland Government	Metallurgy and science management
Richard Arculus from 2008	Lead scientist of ARC/LIEF grant	ANU	Igneous petrology, volcanology; ODP shipboard participant
James Daniell from 2012	James Cook University	James Cook University, Townsville	Marine geoscience including sea-floor mapping
Neville Exon from 2008	ANZIC Program Scientist	ANU	Marine geology and geophysics; ODP co-chief scientist
Stephen Gallagher from 2008	ANZIC Science Committee Chair	University of Melbourne	Biostratigraphy, foraminifera
Chris Hollis from 2008	Chair, NZ IODP	GNS Science, NZ	Palaeoclimate, micropalaeontology
Ian Jackson	ANU representative	ANU	Earth structure
Chris Yeats from 2008	CSIRO representative	CSIRO Exploration & Mining Division, Perth	Hydrothermal systems. ODP and IODP shipboard experience
Annette George from 2012	UWA representative	University of Western Australia, Perth	Geology and tectonic history of sedimentary basins
Before 2013			
Kate Wilson, 2008–2010	Chair	CSIRO; NSW Department of the Environment	Molecular biology and Biotechnology
David Falvey, 2008	ARC representative	ARC, Canberra	Geophysics
Ian Mackinnon, 2008–2009	ARC representative	ARC, Canberra	Engineering
Richard Coleman, 2010–2012	ARC representative	ARC, Canberra	Marine geology
Patrick De Deckker, 2008–2009	ANU representative	ANU	Micropalaeontology
Andrew Roberts, 2010–2012	ANU representative	ANU	Palaeomagnetics
Giuseppe Cortese, 2010	New Zealand representative	GNS Science, NZ	Palaeoclimate/ palaeoceanography radiolarians

Member	Position	Institution	Expertise
Will Howard, 2008–2009	University of Tasmania representative	University of Tasmania, Hobart	Palaeoceanography
Kelsie Dadd, 2008–2010	Macquarie University representative	Macquarie University, Sydney	Physical volcanology
Janet Hergt, 2008–2010	Melbourne University representative	University of Melbourne	Petrologist
Moyra Wilson, 2010–2011	Curtin University representative	Curtin University of Technology, Perth	Sedimentology, petroleum geology
Paul Dirks, 2010	James Cook University representative	James Cook University, Townsville	Structural geology, geodynamics, cratons
Robin Beaman, 2011	James Cook University representative	James Cook University, Cairns	Marine geoscience including sea-floor mapping

ANZIC Science Committee

The ANZIC Science Committee encourages and assists the development of science proposals, helps organise topical workshops, assesses cruise applicants and applicants for IODP panel membership, and helps get quality speakers to visit Australian and New Zealand research centres. A sub-committee assesses applications for post-cruise scientific funding. Stephen Gallagher has been a thoughtful and conscientious Chair of this ANZIC Science Committee. The membership balanced expertise and regional representation, and members rotated on and off (Table 5.4).

Table 5.4. ANZIC Science Committee: 2008–2013

People	Institutions	Expertise	Years involved
Will Howard (Chair)	University of Tasmania	Palaeoceanography	2008
Stephen Gallagher (Chair)	University of Melbourne	Micropalaeontology (forams); wide experience in science of drill cores	2008–2013
Neville Exon (Program Scientist, Australian IODP Office)	ANU	Marine geology and geophysics; ODP co-chief scientist	2008–2013
Linda Blackall	AIMS	Microbiology	2008–2011

People	Institutions	Expertise	Years involved
Mike Coffin	Director, IMAS, Hobart	Geophysics; ODP co-chief scientist	2012–2013
Lindsay Collins	Curtin University of Technology	Sedimentologist, petroleum geologist	2008–2013
Leonid Danyushevsky	University of Tasmania	Igneous petrologist, oceanic crust, neotectonics	2008–2011
Trevor Falloon	University of Tasmania	Igneous petrologist, oceanic crust; ODP experience	2012–2013
Michael Gagan	ANU	Isotope palaeoclimatology; IODP shore-based participant	2008–2013
Stuart Henrys	GNS Science, NZ	Basin studies, marine geophysics, structural geology	2008–2011
Gary Huftile	Queensland University of Technology	Active tectonics, structural geology; IODP shipboard scientist	2008–2012
Anna Kaksonen	CSIRO Land and Water Division, Perth	Microbiology; ODP shipboard participant	2008–2013
Janice Lough	AIMS	Coral reef studies	2012–2013
Jill Lynch	University of Melbourne	Microbiology; IODP shipboard participant	2011–2012
Helen McGregor	ANSTO and University of Wollongong	Palaeoceanography	2008–2013
Robert McKay	Victoria University of Wellington	Sedimentology, Antarctic glacial history; IODP shipboard participant; ANDRILL	2012–2013
John Moreau	University of Melbourne	Microbiology; IODP shipboard participant	2008–2011
Louis Moresi	Monash University	Plate kinematics	2008–2011
Wouter Schellart	Monash University	Plate kinematics	2012–2013
Jody Webster	University of Sydney	Carbonate sedimentologist; IODP co-chief scientist	2008–2013
Gary Wilson	University of Otago, NZ	Palaeomagnetism, Antarctica; Marine drilling expert	2008–2013

Science participation

Selection of participants for expeditions went through two phases. In response to information from IODP about forthcoming expeditions, the ANZIC IODP Office called for applications in a standard IODP-wide format. These applications were considered by the ANZIC Science Committee and suitable ANZIC applicants were ranked. Our rankings and the applications were then sent to those planning each expedition, and the planners built a suitable shipboard scientific team from applicants from around the world, normally including at least one Australasian on each expedition (the ANZIC quota). The participants are listed in Table 5.5.

Many Australians and New Zealanders have been involved in helping to write proposals for work in our region and elsewhere. Lead proponents of accepted proposals have been:

- Jody Webster (University of Sydney): Great Barrier Reef Environmental Change Expedition 325. Drilled in 2010 with Dr Webster as co-chief scientist.
- Richard Arculus (ANU): Izu-Bonin-Mariana Arc Origins Expedition 351. Carried out in 2015 with Professor Arculus as co-chief scientist.
- Stephen Gallagher (University of Melbourne): Indonesian Throughflow Expedition. This expedition to Australia's Northwest Shelf was carried out in 2015.

In 2014, another ANZIC lead proponent of a highly ranked regional proposal was:

- Robert McKay (Victoria University of Wellington): Antarctic Cenozoic Palaeoclimate Proposal 751.

In 2009, ANZIC Governing Council set aside money from scientific members to support post-cruise research by shipboard participants, and this was used very widely.

Table 5.5. ANZIC participants on IODP expeditions: 2007–2013

Expedition	Date	Participants
Chikyu NanTroSeize 1; 316, Nankai Trough Faulting	Dec 2007–Feb 2008	Chris Fergusson (University of Wollongong), sedimentology
Canterbury Basin; 317, sea-level fluctuations in last 20 million years	4 Nov 2009–4 Jan 2010	Bob Carter* (James Cook University), Simon George* (Macquarie University), Greg Browne and Martin Crundwell (GNS Science), Kirsty Tinto (University of Otago)
Wilkes Land; 318, climate and oceanographic changes in last 53 million years	4 Jan–9 Mar 2010	Kevin Welsh* (University of Queensland) and Robert McKay (Victoria University of Wellington), both sedimentology
Chikyu NanTroSeize 2; 319 Nankai Trough deep observatory	10 May–31 Aug 2009	Gary Huftile (Queensland University of Technology), structural geology
PEAT 1; 320, Eastern Pacific environments	5 Mar–5 May 2009	Christian Ohneiser (University of Otago), palaeomagnetism
Chikyu NanTroSeize 2; 322, Nankai Trough Subduction	5 Sep–10 Oct 2009	John Moreau* (University of Melbourne), microbiology
Bering Sea; 323, connections from Pacific to Arctic	5 July–4 Sept 2009	Kelsie Dadd* (Macquarie University), sedimentology of volcanic ash
Shatsky Rise; 324, volcanic buildup in northwest Pacific	4 Sept–4 Nov 2009	David Murphy* (Queensland University of Technology), petrology of volcanics
Great Barrier Reef; 325, environmental change caused by post-glacial sea-level rise (*Greatship Maya*)	11 Jan–5 Mar 2010	Jody Webster* (University of Sydney), co-chief scientist, reef formation. Post-cruise science party Michael Gagan (ANU), palaeoclimate, and Tezer Esat* (ANSTO/ANU), climate
South Pacific oceanic gyre microbiology; 329, east of New Zealand	8 Oct–12 Dec 2010	Jill Lynch* (University of Melbourne), microbiology
Louisville Seamount Trail geodynamics; 330, southeast of Tonga	12 Dec 2010–11 Feb 2011	Ben Cohen* (University of Queensland), volcanic petrology, David Buchs* (ANU), volcanic sedimentology
Chikyu Deep Hot Biosphere; 331, Okinawa Trough	1 Sept–3 Oct 2010	Chris Yeats (CSIRO), sulphide petrology
Costa Rica Seismogenesis Project (CRISP); 334, west of Central America	16 Mar–17 Apr 2011	Gary Huftile (Queensland University of Technology), structural geology
Superfast Spreading Rate Crust 4: 335, eastern Pacific	17 Apr–20 May 2011	Graham Baines (University of Adelaide), petrology

Expedition	Date	Participants
Chikyu Coalbed Biosphere off Shimokita; 337	6 July–15 Sept 2012	Rita Susilawati* (University of Queensland), coal geologist
Chikyu NanTroSeize Plate Boundary Deep Riser; 338	25 Nov 2012–13 Jan 2013	Lionel Esteban* (CSIRO), structural geology/sedimentology/physical properties
Mediterranean Outflow; 339	17 Nov 2011–17 Jan 2012	Craig Sloss* (Queensland University of Technology), sedimentology
Lesser Antilles volcanism and landslides; 340	3 Mar–17 Apr 2012	Martin Jutzeler* (University of Otago), structural geology
Southern Alaska Margin tectonics, climate and sedimentation; 341	29 May–29 July 2013	Christopher Moy* (University of Otago), sedimentology; Maureen Davies* (ANU), physical properties; Carol Larson (NZ National Aquarium), outreach
Palaeogene Newfoundland sediment drifts; 342	2 June–11 Aug 2012	Brad Opdyke* (ANU), palaeoceanography; Chris Hollis* (GNS Science), radiolarians
Chikyu Japan Trench Fast Drilling Project; 343	1 Apr–21May 2012	Virginia Toy (University of Otago), structural geology
Costa Rica Seismogenesis Project (CRISP 2); 344	23 Oct–11 Dec 2012	Alan Baxter* (University of New England), nannofossils
Hess Deep Plutonic Crust; 345	11 Dec 2012–13 Feb 2013	Trevor Falloon* (University of Tasmania), igneous petrologist
Asian Monsoon (Japan Sea); 346	29 July–28 Sept 2013	Stephen Gallagher* (University of Melbourne), foraminifera
Chikyu NanTroSeize Deep Riser Observatory 3; 348	25 Nov 2013–10 Jan 2014	Matthew Josh (CSIRO), physical properties/downhole logging

ANZIC shipboard participants and science party are listed by year (those outside ANZIC's normal quota are in brackets).

All expeditions used the *JOIDES Resolution* unless specified otherwise.

* Post-cruise funding was granted.

There were 36 participants in all: 34 shipboard scientists, including Carol Larson in an outreach role, and two land-based members of a science party, Michael Gagan and Tezer Esat.

- 2008: Chris Fergusson
- 2009: Christian Ohneiser, Kelsie Dadd, David Murphy, Gary Huftile, John Moreau, Bob Carter, Simon George, (Greg Browne, Martin Crundwell, Kirsty Tinto)
- 2010: Jody Webster, Mike Gagan, Tezer Esat, Kevin Welsh, Rob McKay, Chris Yeats, Jill Lynch
- 2011: Ben Cohen, David Buchs, Gary Huftile, Graham Baines, Craig Sloss

- 2012: Martin Jutzeler, Virginia Toy, Brad Opdyke, Rita Susilawati, Lionel Esteban, Alan Baxter, (Chris Hollis)
- 2013: Trevor Falloon, Christopher Moy, (Maureen Davies, Carol Larson), Stephen Gallagher, Matthew Josh.

Special scientific funding

A completely new post-cruise science funding scheme was established in 2012, to encourage Australian scientists to address interesting problems that could be solved by working on legacy material (DSDP–ODP–IODP) and hence increase the output from our overall investments in ocean drilling. Thirteen of the funding applications were regarded as of high standard by reviewers and thus supported (Table 5.6), and the successful groups were allocated up to AU$25,000 each.

Table 5.6. Special ANZIC analytical funding in 2012 for work on legacy material

People and Institution	Title of proposal
David Heslop and Andrew Roberts (ANU)	Searching for giant magnetofossils in the geological record (an integrated analysis of Legs 113, 115, 119, 143, 198 and 208)
Alexandra Abrajevitch (ANU)	Magnetostratigraphy and rock magnetism of Site 747A, ODP Leg 120
Taryn Noble (University of Tasmania) and Michael Elwood (ANU)	Southern Ocean's role in moderating glacial–interglacial variability in atmospheric pCO2: Decoupling nutrient cycling and ocean circulation (Chatham Rise: DSDP 594B)
Gregg Webb (University of Queensland), Luke Nothdurft (Queensland University of Technology) and Jody Webster (University of Sydney)	Tahiti (IODP Leg 310) as a natural laboratory for studying coral and microbialite diagenesis
Frances Jenner (Carnegie Institute) and Richard Arculus (ANU)	Trace element and volatile abundance systematics in the world's largest intra-oceanic igneous provinces
Mark Kendrick (University of Melbourne), Masahiko Honda (ANU) and Richard Arculus (ANU)	Recycled and primitive halogens in backarc basins: Constraints from high precision Cl, Br, and I analyses of basaltic glass (ODP Leg 135: Lau Basin)
Helen McGregor (University of Wollongong)	Great Barrier Reef temperatures and palaeoclimate from the LGM to present: Additional samples from IODP Expedition 325: 'Great Barrier Reef environmental changes'

People and Institution	Title of proposal
Barbara Wagstaff, Stephen Gallagher and Guy Holdgate (University of Melbourne)	A 6-million-year orbital scale record of the onset and variability of the Australian Monsoon: Pollen evidence from ODP Site 765 northwest Australia
Zanna Chase and Taryn Noble (University of Tasmania)	Links between Australian dust and marine productivity (ODP Leg 189: The Tasmanian Seaway and DSDP 593: Challenger Plateau)
Jon Woodhead and Janet Hergt (University of Melbourne)	Constraints on the global subduction flux: State-of-the-art analytical approaches to determining the composition of the altered oceanic crust
Masahiko Honda (ANU), Mark Kendrick (University of Melbourne) and Richard Arculus (ANU)	Noble gas systematics in basaltic glasses from the Lau Backarc Basin: Characterisation of mantle sources, magmatic degassing and crustal contamination
Andrew McNeill, Trevor Falloon, Sandrin Feig and David Green (University of Tasmania)	Sulphur and metal evolution in parental mid-ocean ridge basalt magmas
Amy Chen and Paul Hesse (Macquarie University)	A multi-proxy approach to address water column oxygenation change at the Palaeocene-Eocene Boundary (New Jersey margin)

6

Tales from the ship

These stories from participants aboard a drilling vessel give an impression of life at sea for a couple of months, with a mixture of day-to-day life, background science and the scientific work done by each person. Most had never been on a drill ship before. The authors were not constrained by requirements to stay within a format, and I'm sure readers will be impressed by the variety and the quality of the articles. These articles were first written in 2015, and all deal with the first plase of IODP (IODP(1)).

>>

Shallow megasplay and frontal thrusts: The Nankai Trough Seismogenic Zone Expedition 316 on the *D/V Chikyu*

Chris Fergusson, University of Wollongong

In late November 2007, an informal invitation was issued to Australian IODP from the Japanese Ministry for an Australian scientist to participate in the six-week Expedition 316 to the Nankai Trough off southeast Japan, starting in mid-December that year. This was in response to Australia's successful bid for funding to join the Integrated Ocean Drilling Program, which was planned to begin in January 2008. Without a participant in Expedition 316, it was possible that Australia would not fill all the scientist positions to which it was entitled over the next couple of years. At short notice I applied to go as I had participated in the Nankai Trough Leg 190 of the Ocean Drilling Program. Scientists were to be flown out to the *Chikyu* by helicopter from a base on the eastern Kii Peninsula (over 300 km west-southwest of Tokyo). We were allowed to take no more than 10 kilograms of luggage on the helicopter – so I travelled very light.

The helicopter trip also meant that it was necessary to undertake Helicopter Underwater Escape Training (HUET). HUET was a day-long course conducted at the Japan Agency for Marine-Earth Science and Technology (JAMSTEC) facilities near Yokohama. This involved a lecture and practical component that prepared one for the unfortunate circumstance where the helicopter has to land on or fall into the ocean, where almost invariably it tips upside down. The chances of escaping an overturned inundated helicopter are close to zero without HUET. This proved to be a good bonding experience for the 12 members of the science party who attended my particular training session. Interestingly, the HUET training was conducted by an Australian, a former member of the Royal Australian Navy, whose expertise in safety was being widely utilised by JAMSTEC. In the practical component, we were required to successfully escape from a mock helicopter tilted upside down in an indoor heated swimming pool (remember we were in the Northern Hemisphere in winter). We re-enacted four crash scenarios with increasing difficulty to reflect real conditions. Fortunately, if someone was unsuccessful in escaping during training, they were rescued by trained staff and not allowed to actually drown.

On 19 December, we were flown out to the *Chikyu*. I remember seeing a large pod of dolphins along the way, and the view of the ship from the air was truly impressive. Each person on the ship has their own cabin with adjoining bathroom. Laboratories are spacious and very well equipped, including an X-ray Computed Tomography (X-CT) scanner. This proved very useful as a large number of whole round cores were collected and the X-CT scans provided a means of determining whether valuable data was being lost in whole round samples that were to be used destructively for microbiological purposes.

For me, having spent years mapping turbidites (sediments laid down by turbidity flows running downslope) in orogenic belts, nothing could ever match the thrill of watching Neogene turbidites being recovered from the deep sea. Downhole geophysical logging on non-cored sequences has reached a great degree of sophistication, but it was clear from Expedition 316 that core samples are needed to better understand what is happening in the subsurface. For two sites, we had access to excellent logging results from the earlier Expedition 314; as valuable as those results were, a much clearer picture emerged from the collection of cores. The most exciting part of Expedition 316 was drilling through the structurally complicated, imbricated, frontal thrust at the toe of the Nankai Trough Accretionary Prism (Sites C0006 and C0007).

I was one of five sedimentologists working on the cores in the science party; there were two sedimentologists logging core on each 12-hour shift. My shift was with Uisdean Nicholson, then a PhD student at the University of Aberdeen. On the opposite shift were Kitty Milliken (University of Texas at Austin) and Michi Strasser (then University of Bremen, now University of Innsbruck). Another sedimentologist on the ship was Arito Sakaguchi (JAMSTEC), who worked on the X-CT scanner. I also spent a lot of time in the lab with Xixi Zhao (palaeomagnetism, University of Santa Cruz), Robert Harris (downhole specialist, Oregon State University) and Asuka Yamaguchi (structural geologist, University of Toyko).

A highlight of the cruise was the table tennis tournament. After many exciting matches, it came down to a grand final between Xixi and Michi, which Xixi won convincingly. We were also treated to elaborate festivities at Christmas and New Year, including 'Secret Santa' present-giving where each scientist brought along a gift for someone else. I received a pocket-sized tourist guide to Paris.

The ultimate aim of this drilling was to examine the causes and controls of earthquakes along subduction zones. In this case, the segment of the subduction zone thrust extending 350 km to the west-southwest from offshore Tokyo has been highlighted as playing a critical role in earthquakes and devastating tsunami, such as the 1944 Tonankai event (moment magnitude 8.1), which generated coastal waves up to 10 m high. Understanding earthquakes and tsunami is clearly a major societal priority for Japan. Drilling into an active subduction zone is technically challenging but scientifically rewarding. Ultimately, riser drilling using the *Chikyu* is planned to reach the subduction zone thrust at a depth of 4,600 to 5,000 m below the sea floor, and has already reached a depth of just over 3,000 m below sea floor in Expedition 348.

Post-cruise activities included a conference in Kyoto where results were discussed for the three expeditions (314, 315 and 316) involved in the Kii Penisula transect. Expedition 316 was an earlier part of ongoing deep-sea drilling in the Nankai Trough (Nankai Trough Seismogenic Zone Experiment – Kii Peninsula transect).

Figure 6.1. Christmas Day 2007 on the *Chikyu* with co-chief scientist Liz Screaton (left) and sedimentologists Michi Strasser, Kitty Milliken and Uisdean Nicholson (left to right)

Source: Photograph taken by Chris Fergusson onboard the *Chikyu*, 25 December 2007

Figure 6.2. New Year celebrations in the Core Processing Laboratory, just before midnight, 31 December 2007, on the *Chikyu*

From left to right, scientists in foreground are: Matt Knuth, Daniel Curewitz, Chun-Feng Li with back to camera, France Girault and Fred Chester.

Source: Photograph taken by Chris Fergusson onboard the *Chikyu*, 31 December 2007

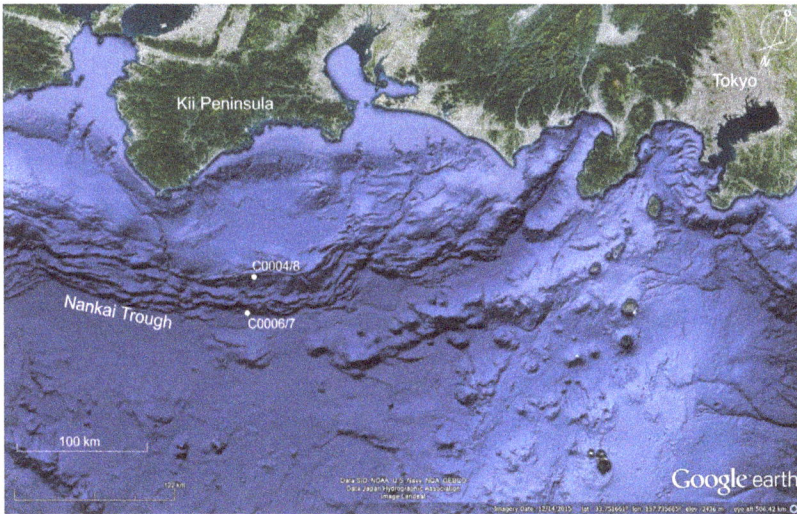

Figure 6.3. Google Earth image showing the locations of Expedition 316 drill sites

Sites C0004/C0008 target an upper trench-slope basin in the footwall of a major splay fault off the subduction zone megathrust. Sites C0006/C0007 drilled down to the main frontal thrust at the seaward edge of the Nankai Trough Accretionary Prism, i.e. the tip of the subduction zone megathrust.

Source: Google Earth image annotated by Chris Fergusson

Tales from Wilkes Land: IODP Expedition 318

Rob McKay, Victoria University of Wellington

Sailing out of my hometown port of Wellington en route to the Antarctic on the *JOIDES Resolution* on IODP Expedition 318 was one of the proudest and most exciting moments of my life. I have been incredibly lucky with timing and opportunities in my career, and IODP has been instrumental in a large part of my early career successes. I have always thought you make your own luck, but this is not entirely true in my involvement in IODP. I was in the final stages of my PhD at Victoria University of Wellington working on sediments recovered as part of the ANDRILL project – another drilling project in the Antarctic that is highly complementary to the IODP program – when I first saw that IODP was planning to drill on the Wilkes Land margin of the East Antarctic Ice Sheet. I knew immediately that here was an amazing opportunity to directly address some of the questions that had developed from our ANDRILL work, and that my skill set would hopefully be deemed useful for the expedition. Most importantly, the drilling was scheduled to be undertaken within a year of my PhD completion, so it was the most logical next step in my career pathway to becoming a research scientist. Not many completing PhD graduates get such an opportunity.

However, this is where the luck, and the hard work of others, was required. I knew my chances of being a shipboard scientist were slim as New Zealand was not a member of IODP at the time. Due to my experience in Antarctic drilling, I thought I stood a good chance of being selected on scientific merit, and I could have tried to look overseas for a postdoctoral position to work on these cores. This was not a preferred option, as I had already done my overseas living experience prior to my PhD and I was nicely settled back into the Kiwi way of life – staying in New Zealand was a high priority for me. It would also be difficult to find a postdoctoral position on this project without a confirmed berth. Fortunately, by the time I had finished my PhD, several New Zealand universities, along with GNS Science, recognised the huge scientific potential that IODP offers and we joined the Australian consortium for IODP membership. This allowed me to immediately send in my application to sail, and I was accepted on the IODP Wilkes Land Expedition 318.

Having been selected to sail, I was able to apply for a highly competitive Foundation for Research, Science and Technology Postdoctoral Fellowship. Being involved in an international collaboration of the scale of IODP, and addressing first order questions about our planet's climate and ocean system, were instrumental factors in my successfully obtaining this fellowship. The fellowship allowed me to work full-time on the cores collected during the Wilkes Land cruise. My involvement in IODP also eventually led to my full-time position as a lecturer at Victoria University, a Marsden Fund grant, a Prime Minister's MacDiarmid Emerging Scientist award, and my current Rutherford Discovery Fellowship award. I can honestly say that I would have almost certainly had to leave New Zealand after my PhD to pursue an academic career working in a similar IODP-related field if I had not sailed on this expedition, and if others had not worked hard behind the scenes to secure membership of IODP.

Big science more often than not requires big investments and large collaborations, and IODP is no exception. I was amazed stepping onto the *JOIDES Resolution* for the first time and seeing her world-class laboratories as well as the technical capabilities she had to drill sediment cores, even in the most hostile of drilling environments in the Antarctic. However, it was encountering the range of expertise and experience of the other scientists onboard that I came to value the most. Being in a closed environment for nine weeks with 30 other like-minded Antarctic scientists from a wide range of sub-disciplines was highly intellectually stimulating. I have not only made a range of powerful scientific collaborations, many of which are proving extremely productive, but I also made friends for life from all corners of the globe.

It is difficult to explain to people who haven't sailed on an IODP expedition the bonding experience that occurs amongst the scientists. The shifts were long and tiring, fresh food ran out halfway through the expedition, there were hurricane-strength winds and giant waves in the Southern Ocean, sea sickness, and we had no days off, not to mention being stuck aboard a 470-foot-long ship for almost 70 days. However, exciting new scientific discoveries each day, the thrill of seeing full drill core barrels being pulled up from the sea floor, a common sense of purpose, engaging and vigorous scientific discussion, not to mention an excellent coffee machine, more than offset the negative aspects. It also helped that we were travelling to one of the most remote and beautiful places on Earth. I had been lucky enough to see the beauty of the Antarctic before then, with several expeditions to the McMurdo Sound region, but I had never sailed down there. Now it is hard to imagine any better way to travel there – the raw power of the roaring forties and furious fifties in the Southern Ocean is something to behold. Seeing it at first hand, it is hard to comprehend that anything could live there, despite it being one of the most biologically productive zones on Earth. Sailing past armadas of giant icebergs, giant waves crashing over the bow of the ship and visits by the occasional humpback whale and penguin are truly awe-inspiring experiences that I will never forget.

I always tried to make it outside in the evenings to take in our unique surroundings and the fresh Southern Ocean air. The morale was always high onboard, in part due to the beauty of the Antarctic environment, but also due to the quality of the science being conducted. During the cruise, we felt we were doing something pioneering, and this didn't end when we sailed into Hobart after nine weeks at sea. Collaborations are still continuing, and I have recently had incredible opportunities to visit and work on IODP material in labs in the US, UK and Spain, undertaking studies with colleagues who are truly world leaders in their respective fields. The quality of the science and the publications that have already resulted from this expedition are exceptional, even by IODP standards, and more is to come. On this note, I truly believe the benefits of my IODP experience will continue throughout my entire career.

Figure 6.4. Crossing the Antarctic Circle with Rob McKay in the front of the group

Source: Rob Dunbar, Stanford University

Figure 6.5. Approaching Antarctica with gentle waves, sea ice and a lonely bird

Source: Rob Dunbar, Stanford University

Figure 6.6. A party of icebergs; remember that 90 per cent of the volume is underwater

Source: Rob Dunbar, Stanford University

Two months near the Pacific Equator on IODP Expedition 320

Christian Ohneiser, University of Otago, Dunedin

I received word that a position as palaeomagnetist had unexpectedly opened on IODP Expedition 320, the 'Pacific Equatorial Age Transect' (PEAT), only a few months before the *JOIDES Resolution* (*JR*) was due to set sail. After reading the proposal, I knew this would be a scientific game changer and great opportunity to be part of an exciting expedition with some of the world's leading minds in palaeoclimatology. The plan was to recover a composite sedimentary succession from the Equatorial Pacific, spanning the Eocene to today, in an effort to understand the influences of tectonic (Panama and Indonesian sea gates closure) and climate transitions on the evolution of the Pacific Ocean. For us palaeomagnetists, the objectives were to help build the chronology at each drill site, to refine the geomagnetic polarity timescale and to build the next generation of relative palaeointensity records for older portions of the Cenozoic.

Having arrived in Hawaii I soaked up the sun, enjoyed some surf and, most importantly, I consciously enjoyed the feeling of land beneath my feet before setting off for two months at sea. A few days later, as the gangway was raised onto the *JR*, I was told by one of the ODP veterans that this was the worst moment during the expedition; there was no going back now. We settled into our shifts; mine was midnight to midday, which I dreaded because I am by no means a 'morning person'! First core arrived on deck a few days later, and the routine measurements and core characterisation began.

Palaeomagnetic measurements and demagnetisation of half cores were conducted using the shipboard superconducting magnetometer; I was familiar with the instrument but not the new software. The first site (U1331) targeted the oldest sediments of the expedition and palaeomagnetic data were flowing quickly from the magnetometer. Demagnetisation data were easy to interpret, and working side-by-side with experienced micropalaeontologists was a real pleasure. The micropalaeontologists provided preliminary ages within minutes of cores landing on deck, which meant that in a matter of hours the first geomagnetic reversals were identified and correlated and the age models began to take shape. It was a palaeomagnetist's dream!

Figure 6.7. Christian Ohneiser about to take small cubic palaeomagnetic samples from a sediment core
Source: Rob Dunbar, Stanford University

As the days melted into weeks, we settled into our routines. Breakfast was at 2330 hours before taking over from the off-going crew. We quickly formed a tight bond with loosely set traditions, such as a 0300 hours espresso break on deck watching squid and flying fish that were drawn to the *JR*'s lights. Second breakfast usually came at 0600 hours, which was preceded by a sunrise coffee. Within a few weeks, we concluded that our favourite time of day was between 0100 and 0500 hours, because after 0600 hours things usually became hectic with the arrival of a bridging shift, the technicians and finally the arrival of curious day-shifters. At 1200 hours we knocked off and, depending on what mood we were in, we went to the gym, sunbathed or watched one of the numerous antique laser disc films on offer in the boat library. My personal favourite pastime was jogging on the helideck. The helideck was my bright green and yellow place of solitude under the equatorial sun with the occasional smell of diesel fumes from the smoke stacks. However, because the helideck was quite small, a lap was finished in under a minute, which meant that after 10 laps I was forced to run in the opposite direction.

As the expedition progressed, data amassed quickly. Coring sites were usually less than 24 hours steaming apart, meaning that as each coring site was completed and we began preparing the initial site report (no small task), new cores were arriving on deck and required measuring. Grey hairs began to appear and stress levels grew as the backlog of work increased and Week Six was upon us. We had now cored, measured and logged an almost complete Eocene succession, the entire Oligocene and were getting stuck into the later part of the Miocene. The expedition was already an astounding success with pristine palaeomagnetic records in ~80 per cent of recovered sediments and age models; the age models would change little after the cruise. Oddly enough, we recovered a turbidite from the deep Pacific at Site 1331 and a repeated section at Site 1332, where we recovered the Eocene–Oligocene boundary six times in three holes!

On 26 April, in the dead of night, the last core arrived on deck. The drill string was brought back on deck and the *JR* began its long return journey to Hawaii. We completed our final edited versions of the preliminary site reports as we rounded the southern tip of the Island of Hawaii; the first time we'd seen land in eight weeks. Twenty-four hours later, we docked in Honolulu where the US border police were waiting to deport me because of a minor visa technicality. Because of my late entry onto the expedition, I had not applied for a business visa. Multiple people assured me that the standard visa issued under the visa waiver program was sufficient; I would be on US soil for only a few days before the expedition and would be issued a new visa upon my return. However, because the expedition was returning to a US port, this was classed as one stay – clearly I was an illegal alien in need of deportation! I was handed the 'detain and deport' documents within minutes of the border police boarding the *JR*. Captain

Alex warned me in his thick Scottish accent that 'they don't have fluffy handcuffs in the US' and my dreams of a few laid-back days in Hawaii diminished. The Hawaiian border police were an understanding bunch though, and issued me with an emergency B1 business visa; I had used my only get-out-of-jail-free card in the state of Hawaii. We stepped onto land and found the nearest watering hole, where we sat in the sun and immediately began retelling our war stories from the preceding weeks.

Scientifically, Expedition 320 was an astounding success. We recovered over 3,500 m of sediment spanning the earliest Eocene to recent times and the palaeomagnetism team recognised and correlated over 800 geomagnetic reversals; no mean feat in anyone's book. However, PEAT was not yet complete; another crew boarded the *JR* to core the remaining sites.

A year later, we met in Paris for the post-expedition workshop at the Pierre et Marie Curie University. We presented the work we'd done and cemented the partnerships and collaborations we had promised one another at the conclusion of the expedition. The palaeomagnetists have since developed the first high-quality Remanent Palaeomagnetic Intensity records spanning from the Miocene to the Eocene. These are the foundations for the next generation of palaeomagnetic dating tools, which will allow us to build even more precise age models and correlate successions from around the world. The first comprehensive and continuous record of the evolution of the calcite compensation depth in the Pacific has since been published in *Nature* (see Chapter 13). Today, work continues on the construction of stable isotope records for the entire composite succession, which will provide a new understanding of the evolution of the Pacific Ocean during the Cenozoic. Watch this space for more exciting results from PEAT.

Bering Sea Palaeoceanography: IODP Expedition 323

Kelsie Dadd, Macquarie University, Sydney

The Bering Sea, a marginal sea in the North Pacific, located north of the Aleutian Islands, was targeted to investigate the evolution of global climate on timescales from millennial to Milankovitch during the Plio-Pleistocene.

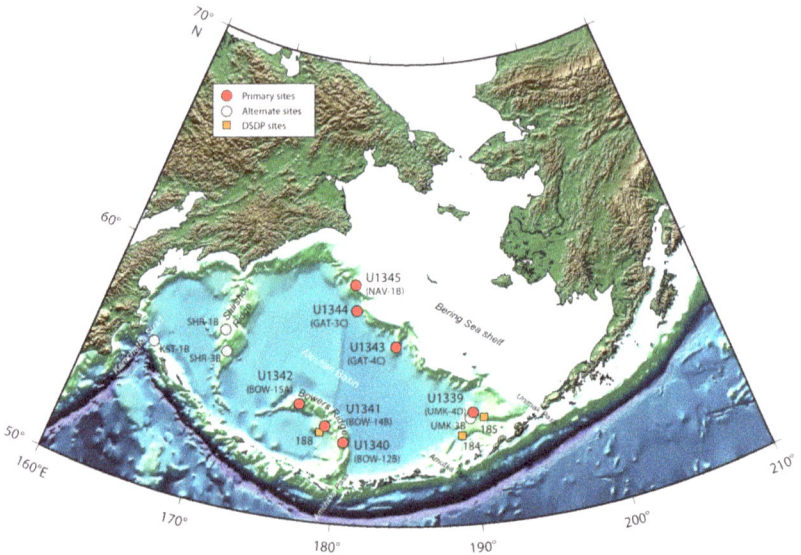

Figure 6.8. Location of the Bering Sea showing the sites drilled on Expedition 323 (red dots)

Source: Expedition 323 Scientists, 2010. Bering Sea palaeoceanography: Pliocene–Pleistocene palaeoceanography and climate history of the Bering Sea. *IODP Preliminary Report* 323. doi.org/10.2204/iodp.pr.323.2010

The basin is important for global climate as it is a site of intermediate water generation. In addition, the evolution of the Bering Sea gateway has almost certainly had an effect on global climate. During the expedition, 5,741 m of sediment (97.4 per cent recovery) was drilled at seven sites, with holes up to 745 m below sea floor and spanning 5 million years ago (Ma) in age. I sailed as a sedimentologist, and described and sampled core as fast as it was landing on the catwalk.

Figure 6.9. Drill core on deck. There was so much expanding gas in the sediment that it shot out of the core barrel when it arrived on deck

Source: William Crawford, IODP/TAMU

Highlights of the expedition for me included seeing the changes in climate and the extent of sea ice recorded in the amounts of ice-rafted debris and sediment–diatom ratio, and also the fluctuations in the oxygen content of the bottom water as reflected in the preservation of delicate laminations (low oxygen) or intense bioturbation of the sediment (high oxygen) (Figure 6.10).

Figure 6.10. Alternating laminated and bioturbated sediment in Site U1432

Source: Kelsie Dadd

I sampled the ice-rafted debris over 2 mm in size, recording the shape and composition of 136 clasts. Ice-rafted debris are particles that have been carried out to sea by ice to be later deposited when the ice melts, and their presence can indicate cooler climates throughout geological history. Many of the clasts were rounded and likely spent time on a beach being moved by wave action when there was little ice cover. They were later picked up by sea ice and eventually dropped into the bottom sediments as the ice melted (Figure 6.11).

Figure 6.11. Rounded dropstone in mud at Site U1341
Source: Tatsuhiko Sakamoto, JAMSTEC

The Bering Sea is located behind the Aleutian volcanic arc and adjacent to the Kamchatka arc and Alaska. All these volcanic provinces have been active over the time interval recovered in cores from the Bering Sea. This volcanic activity is recorded in the sediment of the marginal basin with the frequency, distribution and thickness of ash layers dependent on factors such as the size of the eruption, type of eruption and wind direction. Ash layers can be related to individual eruptions via dating and chemical matching. Ash from 69 ash beds has been analysed for major elements and shows that there are a range of sources from the Aleutian Islands and Alaska.

This was my first IODP expedition and I loved it. I made research contacts from around the world and also made close friends. I particularly liked having the opportunity to learn from the other scientists. While I was familiar with terrigenous sediments and their composition, I had never worked with biogenic sediments and this was a fairly steep learning curve, but one that my fellow sedimentologists and the palaeontologists were happy to help me with. I can now identify a whole range of microfossils in smear slides. I also learnt a lot about sediment geochemistry, palaeomagnetism and climate change over the last 5 million years, and shared my knowledge of volcanism and rock geochemistry with others.

Figure 6.12. Kelsie Dadd on the *JOIDES Resolution*
Source: William Crawford, IODP/TAMU

We sailed from Victoria, Canada, to Yokohama, Japan, but between ports saw almost no land. We did see an amazing amount of sea life and were often accompanied by Dall's porpoises, kittiwakes and fulmars, but also saw fur seals, Pacific dolphins, orcas, a minke whale and a fin whale. The original planning for the expedition included sites on the Shirshov Ridge, a submarine ridge basement high in Russian waters. However, Russia never gave us permission to drill. We sailed with two empty berths, set aside for Russian observers. To end the expedition, we sailed in front of a cyclone from the Kamchatka Strait to Japan, leading to a very bumpy ride back to land.

A study of a huge volcanic plateau: Life as an alteration petrologist aboard IODP Expedition 324

David Murphy, Queensland University of Technology, Brisbane

I arrived in Tokyo's Narita airport in late 2009 to begin a two-month-long adventure to investigate a submerged Jurassic plateau, the Shatsky Rise east of Japan, on IODP Expedition 324. My first introduction to the IODP was a shared ride to the *JOIDES Resolution* with IODP veteran Jim Natland. While the sights of Tokyo flashed by, Jim and I started out the trip by getting stuck into the hypotheses of the expedition – do oceanic plateaux form by mantle plume or plate processes? So continued two months on the ship where science was the main focus and all else was secondary.

The trip started out pretty easily, with the first five days in port in Yokohama. I met and began to get to know the science crew that I'd be working with, none of whom I had met before. We were shown around the living area and inducted into the labs onboard, and started to work together to develop a plan for what we hoped to achieve once drilling started. In the evenings, our Japanese colleagues had some fun showing the first-time visitors to Japan the sights of Yokohama and a little of Japanese culture and nightlife. I am still unsure as to what exactly I ate, but it was mostly delicious. The fun in Yokohama finished far too soon and we sailed from port into the open Pacific – we would not see land again for two months.

The first drill hole was when the proper work finally began. We were aiming for the summit of the Shirshov Massif, the northernmost and smallest of the three edifices of the Shatsky Rise. The growing excitement of the science crew was almost overwhelming when the first drill core was brought on deck. This bubble soon burst when the first sediment cored on the expedition turned out to be a 12 cm piece of chert, which was not what the dominantly igneous petrologists of the science crew were interested in. A few cores later, we finally got some basaltic rocks and were all far happier. From that point onwards there was little let up.

Days soon became a routine of detailed core logging, sample selection for onboard analysis of both petrology and geochemistry and, almost straight away, report writing. We worked in 12-hour shifts, often working almost simultaneously on the newly arrived cores, the petrology of cores from two or three days ago, and report writing for core that was drilled four or five days previously. Also, in the back of your mind, you are thinking about what science you would like to do after the cruise, how the drill core fits with your pre-trip science plan and what samples you would like to take.

What an experience! I don't normally get seasick, but staring down a petrographic microscope in a 5 m swell from Typhoon Choi-Wan, motion sickness was thrown up to a whole new level. The combination of the swirling colours of a rotating thin section in cross-polarised light, the effort to stop my eyes hitting the eyepieces and the rolling and rocking motion of the ship was too much. I needed a break from the stack of thin sections that all needed to be described. It had seemed so important at the sampling meeting, when the night shift handed over to the day shift, to get thin sections made from every single volcanic interval that was drilled two days before.

I had a choice of a short break to stop my head spinning, or a change of task to log yesterday's cores look at the geochemical data from previous cores or work on a site report. Actually, I went for a breather and had a chat with some crewmates from all around the world about the trip so far, about life onboard the ship so far and about what other projects we were all working on at the time.

There were times onboard when the ship's isolation became very obvious. I was certainly a long way from home and a long way from my young family. The inability to go for a decent stroll or to play with my kids and partner got to me. But it was good to be able to discuss normal everyday life with my colleagues and listen to their perspectives – to try, with the others onboard, to get a balance between doing the job and getting your mind off the job.

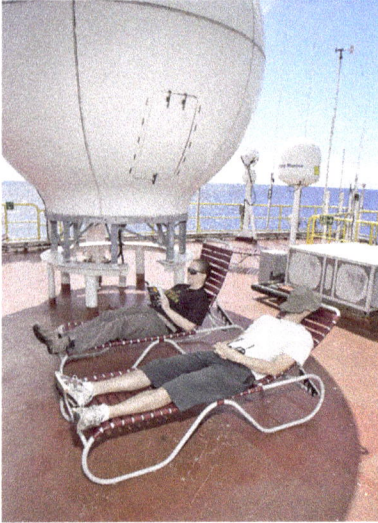

Figure 6.13. David Murphy
(foreground) relaxing aboard ship
Source: David Murphy

Nevertheless, the bond that we developed through working together as a team to produce the raw data and the field reports, and the shared life experience of living onboard the *JOIDES* in isolation from the rest of the world, is something I will always treasure. We all got through the trip just fine, and we got some fabulous rocks to investigate. I am still working with a number of the Expedition 324 crew, both on continued research into the Shatsky Rise and on projects that we dreamed up while onboard.

The objective of the trip was to find evidence to test the formation of oceanic plateaux: essentially whether they are formed by the impact of a mantle plume head on oceanic crust or purely by shallow plate tectonic processes involving unusually large volumes of volcanism.

Shatsky Rise, the third-largest oceanic plateau preserved on oceanic crust, formed at the Jurassic–Cretaceous boundary (about 145 million years ago) when there were abundant magnetic reversals preserved in the surrounding oceanic crust, which allow us to deduce the tectonic environment in which the plateau formed. Results to date indicate that, as in many scientific questions, there is no black or white answer. Certain evidence supports a plume involvement, including high degrees of partial melting, age-progressive volcanism, isotope evidence of a deep mantle source and evidence that the plateau shoaled. Other evidence indicates a significant role for shallow plate tectonic involvement, most importantly magnetic lineations that indicate Shatsky Rise formed on a triple junction of mid-ocean ridges.

In addition, a significant outcome from Expedition 324 is that the Tamu Massif, the largest of the three edifices that make up the Shatsky Rise, is a single volcanic edifice and could be the largest single volcano on Earth. It is comparable in size to the largest volcano in the Solar System, Olympus Mons on Mars.

Aboard *Chikyu* studying the deeply buried microbes in the sediments being subducted in the Nankai Trough: IODP Expedition 322

John Moreau, University of Melbourne

In September–October 2009, I had the amazing opportunity to serve as shipboard microbiologist for IODP Expedition 322 to the Nankai Trough (southwestern Japan). Although the focus of the expedition involved factors controlling subduction zone seismicity, the expedition was also a fantastic opportunity to take samples from the deep subsurface biosphere and investigate the extent of microbial life in the subducting sea floor. I was very excited to go to sea after spending several days in Yokohama at JAMSTEC practising safety drills and learning about shipboard safety protocols.

Once onboard the drill ship *Chikyu* ('Earth' in Japanese), we quickly settled into our shipboard duties and routines. Before I mention those, however, I should say that *Chikyu* offered pretty luxurious surroundings for ocean drilling research: the quarters, while not spacious, were not shared accommodations; the cafeteria served pretty decent food 'round the clock'; movies could be watched with a group in the movie room; the helideck made an excellent jogging track and, at night, a stargazing platform. The laboratory facility was first-class and the crew were friendly and helpful.

Alongside the science team and the ship's crew, the drilling crew at work were amazing to watch; many of them hailed from drilling vessels and platforms in Europe and Indonesia, with experience drilling in deep waters and rough seas. This experience would turn out to be helpful on our expedition as we had to divert, reroute and delay around three(!) tropical storms. Although we skirted the brunt of it, the result of Tropical Storm Choi-wan was >10 m waves that felt like a rollercoaster. Going up a flight of stairs, you felt like you had lead weights in your shoes, while going down felt as if gravity suddenly didn't exist! Even the most seasoned sailors were reaching for their motion sickness medication. Fortunately, we had an able captain and crew who kept us far from any real danger that storm season.

For my shipboard scientific work, I prepared all of the materials and reagents I would need for sampling and processing of sediment cores and sea-floor basalts, as well as conducting anoxic water analyses of sediment pore fluids. I also served on the geochemistry team, processing pore water samples and analysing them for dissolved sulphide, an indicator of bacterial sulphate reduction. I also processed sediment samples for gas analyses (methane and other natural gases). Part of the reason for this doubling (or tripling) up of shipboard duties was that, unlike many other IODP expeditions that explored the deep subseafloor biosphere, this expedition only carried one microbiologist (myself) and one organic geochemist (Dr Verena Heuer, MARUM, University of Bremen). Thus, we rotated and sampled for each other on 12-hour shifts throughout the cruise, changing at noon to allow us both to get some sunlight!

Figure 6.14. Organic geochemist Verena Heuer and microbiologist John Moreau on *Chikyu*

Source: Marta Torres, Oregon State University

The experience I gained from helping with pore water geochemistry and organic geochemistry analyses really enriched my IODP experience in ways that many of my microbiologist colleagues who served on other expeditions may not have enjoyed. I was allowed to have some input into sampling and analysis planning, for example, and that turned out to be useful. Specifically, when discussing the potential geochemical signatures of possible anaerobic sea-floor microbial processes with my geochemical colleagues, I suggested that we try to sample all sediment pore waters anoxically (using a partially sealed bag flooded with nitrogen gas during handling of each sediment core). At first, they disagreed with the need to take this extra step because it was assumed that the concentrations of dissolved metals like iron and manganese would be negligible due to the penetration of oxic seawater. However, after the shipboard pore water analyses were conducted, the geochemistry team leader (Dr Marta Torres, Oregon State University) told me that they detected significant levels of dissolved manganese and iron that they would have otherwise missed had they not used the nitrogen gas bags. This observation, combined with the sulphide analyses, turned out to form the basis for our observation of the unusual 'upside down' redox gradient across sediments closer to the basement crust that we described in Torres et al. 2015 (*Geobiology*, doi.org/10.1111/gbi.12146). From this observation, we concluded that deeply penetrating and recirculating seawater could drive microbial activity at great depths, specifically the oxidation of methane coupled to sulphate reduction at depths >400 m below the sea floor (mbsf). Previously described sulphate–methane transition zones were all found at ~150 mbsf or less. We had discovered one of the deepest sulphate–methane transition zones found thus far in the deeply buried subducting sediments of the Nankai Trough.

Excited about the prospects of describing a deeply buried and unexpected microbial biosphere, my team set to recovering pristine DNA from sediment cores and basalt samples, once these were received onshore in Australia. We were very lucky to have access to a sampled piece of the actual contact zone between the deepest (oldest) sediment and the shallowest (youngest) sea-floor basalt, where the susceptibility of basalt to seawater and microbial weathering was expected to be very high. However, our initial work on this sample has proven it to be extremely difficult: painstaking care is required to avoid any lab-derived contaminants while also extracting sufficiently high-quality DNA for analysis. We are still (several years later) working on this material. Because of the precious nature of this 'interface' sample, our initial onshore research resulted in a publication focused on shallower sediments. This work (along with some samples from Expedition 329) has formed part of the thesis of a PhD student, Toni Cox, who has been a senior author on a *Geobiology* article on Expedition 322 (see Chapter 13).

Our discovery of two 'ultra-deep' (400+ mbsf) sulphate–methane transition zones in the Nankai Trough has extended our knowledge of the habitability of the deep subseafloor, and linked what is known about deep hydrogeologic circulation of seawater from distant exposed volcanic knolls to the stimulation of microbial activity where it would otherwise not be likely to exist – in subducting sediments receiving seawater from below through the basement crust. This finding has interesting implications for the distribution of microbes and preservation of microbial 'oases' in otherwise nutrient-poor environments.

Searching for subseafloor life: IODP Expedition 331

Chris Yeats, then CSIRO, now Geological Survey
of New South Wales

From 1 September to 3 October 2010, IODP Leg 331 on the drill ship *Chikyu* drilled five sites at the Iheya North hydrothermal ('black smoker') field, in the Central Okinawa Trough back arc basin, south of Japan.

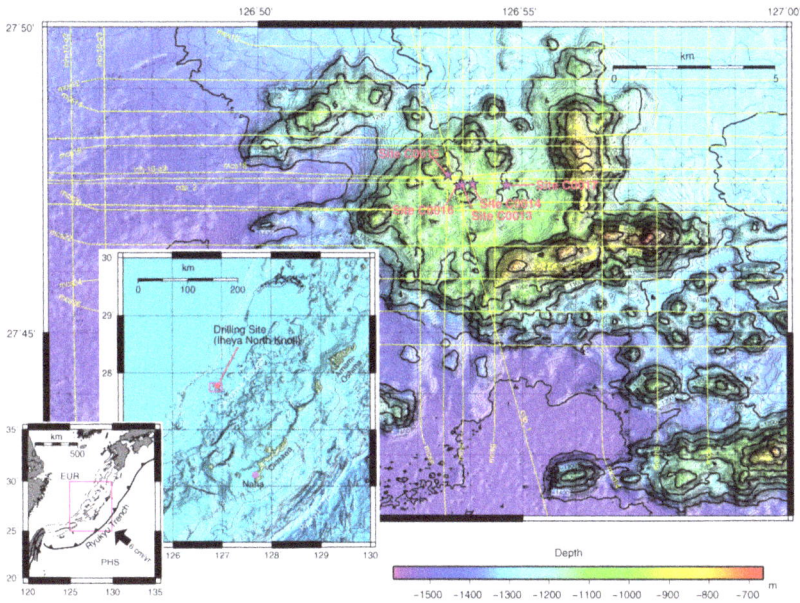

Figure 6.15. Area map of Iheya North Knoll showing Sites C0013–C0017 drilled during Expedition 331

Inserts show the Iheya North Knoll in relation to Okinawa and Okinawa in relation to major tectonic components. EUR = European Plate, PHS = Philippine Sea Plate.

Source: Map sourced from Expedition 331 Scientists (2010). Deep hot biosphere. *IODP Preliminary Report*, 331. doi.org/10.2204/iodp.pr.331.2010

IODP Expedition 331 was unusual for a couple of reasons. First, the 34-day expedition was significantly shorter than a typical two-month IODP expedition. Second, and more significantly, Expedition 331 'Deep Hot Biosphere' was the first IODP expedition, and still one of only a handful of cruises, to focus principally on biology rather than geology. The impact of this focus on the scientific party was profound: membership was dominated by microbiologists, biochemists and fluid chemists, for whom the recovered core was a means not an end. Consequently, logging of the 312 m of drill core recovered during the expedition was left to a small team of three to four geologists, most of whom had been attracted to the cruise by the possibility of drilling metal-rich massive sulphides – but more on that later.

Due to its proximity to the Asian continent, the Okinawa Trough contains significant quantities of organic-rich sediment. We therefore anticipated (and in some cases got) drill core containing levels of hydrogen sulphide (H_2S) that could be harmful to humans, and initial collection of core was conducted wearing breathing apparatus, until it was declared safe. This meant that some members of the science party, including myself, and all of the *Chikyu* core technicians needed to be trained in safe operating procedures for H_2S prior to departing from Shimizu. Wearing heavy overalls, breathing apparatus and a full face mask with temperatures in the mid-30s and humidity close to 100 per cent is a pretty good substitute for sauna therapy!

Previous work by Japanese scientists at Iheya North, including manned and unmanned submersible dives and extensive seismic surveys, had given the cruise participants a fair idea of what to expect at the site. Hydrothermal alteration and massive sulphide mineralisation at Iheya North is hosted in a geologically complex, mixed sequence of coarse pumiceous volcaniclastic and fine hemipelagic sediments (basically gravel and mud), overlying a dacitic to rhyolitic volcanic substrate. The principal aim of the expedition was to sample the hydrothermally altered sediments and prove the existence of a functionally active, metabolically diverse subvent biosphere associated with subseafloor hydrothermal activity in the Iheya North field. This goal wasn't achieved because, fundamentally, things got hot a lot faster than we expected. The initial strategy was to drill using Perspex core liners, but we soon abandoned that strategy at our first site and switched to aluminium when it became obvious that the plastic couldn't survive the temperatures we were encountering.

Subsurface temperatures at the three hydrothermally altered drill sites rapidly exceeded those that could support life, effectively meaning that the biologists were sampling sterile material. However, each of these three sites was successful in drilling into hydrothermal fluid reservoirs and created artificial hydrothermal vents, which have been monitored and sampled by Japanese researchers for the past five years.

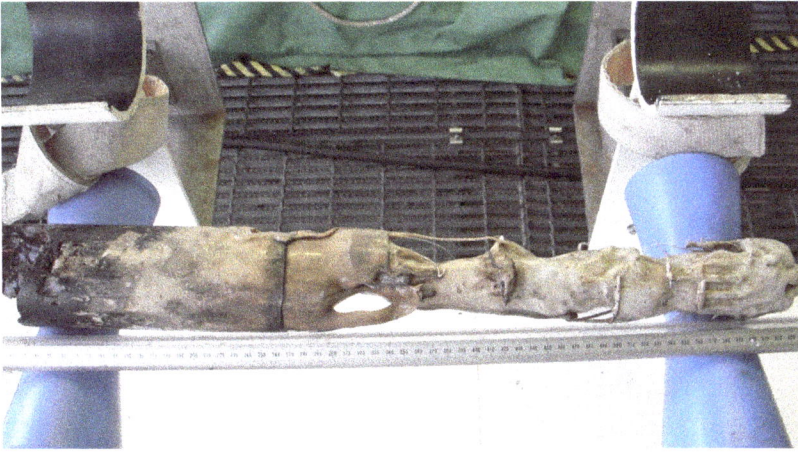

Figure 6.16. All that remained of a 9 m Perspex core liner after an early drilling attempt at our first site, melted by the very high temperature

Source: Photograph taken by Chris Yeats aboard *Chikyu*

As an ore deposit geologist, what excited me most about Leg 331 was the proposal to drill directly into a large actively venting massive sulphide mound – the 20 m high North Big Chimney (NBC) mound, which has measured venting temperatures of 311°C. I had been a shipboard scientist for the Ocean Drilling Project's previous attempt to drill an active hydrothermal field in a backarc basin environment (ODP Leg 193 to the Bismarck Sea, Papua New Guinea, in 2000), but that expedition deliberately sought to avoid drilling known massive sulphides – and was successful in that goal.

Leg 331's first attempt to drill NBC, about midway through the cruise, ended disastrously. The hole was spudded in the top of the mound, but did not drill vertically. Consequently, when the *Chikyu* drillers attempted to withdraw from the hole to recover the core (they were using a non-wireline system adapted from the oil and gas industry), the rods became jammed and ultimately broke. The drill bit remained lodged in the hole and all core was lost. Our second attempt at the very end of the cruise was much more successful. The hole was drilled adjacent to the base of NBC and successfully penetrated 45 m. Unfortunately, after the experience of the first hole at NBC, the decision was made to minimise equipment risk at the site and the hole was cored without a core catcher, meaning that only 1.7 m of core was recovered. However, the material recovered included an interval of black zinc-rich massive sulphide (Figure 6.17), which has the same mineralogy and textures that are seen in Kuroko Black Ore.

Figure 6.17. Zinc-rich massive sulphide drilled during IODP Leg 331
Source: Photograph taken by Chris Yeats aboard *Chikyu*

Associated hydrothermally altered volcanic rocks recovered from further down the hole also resembled those seen associated with this type of mineralisation on land, meaning that this site is an outstanding modern analogue for Kuroko-style mineralisation. This is arguably the most important scientific result to come out of IODP Leg 331.

I have been fortunate enough to participate in over 20 research expeditions during my career on Australian, American (including *JOIDES Resolution*), Japanese, British and Indonesian research vessels. In terms of available analytical facilities and drilling capability, DV *Chikyu* is undoubtedly the most impressive facility I've sailed on. The vessel is enormous and, unique amongst the ships that I've sailed on, incredibly stable under normal weather conditions. It really doesn't feel like you're at sea at all. The crew are friendly and helpful, the food is amazing (but I do love sushi) and the accommodation is first class. This incredible research facility has not been available to IODP as much as could be hoped, but in 2016 new possibilities started to open up. I can only hope that in coming years other Australian scientists will have the opportunity to sail on what was and should be one of the IODP's most valuable assets.

The onset of the Mediterranean Outflow Current and its influence on global climate change: IODP Expedition 339

Craig Sloss, Queensland University of Technology, Brisbane

I joined the IODP Mediterranean Outflow Expedition 339 in late 2011 as a sedimentologist. The two-month expedition recovered 5 km of sediment core samples from an area never before drilled along the Gulf of Cadiz and west of Portugal. My role aboard the research vessel *JOIDES Resolution* was to provide sedimentological and stratigraphic analysis of collected cores. We found new evidence of a deep-earth tectonic pulse, retrieved a detailed record of oceanographic climate changes and even made key discoveries that could assist the future of oil and gas exploration. The Strait of Gibraltar is a most important oceanic gateway, which reopened less than 6 million years ago after being isolated from the Atlantic for several hundred millennia. Today, deep below the sea surface, a powerful cascade of Mediterranean water spills out through the Strait into the Atlantic Ocean. Because this water is saltier than the Atlantic, and therefore heavier, it plunges more than 1,000 m downslope, scouring the rocky sea floor and carving deep-sea canyons. Currents running along the contours build up mountains of mud called contourites. They hold a record of climate change and tectonic activity that spans much of the past 5.3 million years.

Dorrik Stow, expedition co-chief scientist, commented:

> The expedition brought us many of the eagerly anticipated answers to our questions, as well as wholly unexpected scientific results.

Javier Hernandez-Molina, the second co-chief scientist, commented:

> We set out to understand how the Strait of Gibraltar acted first as a barrier and then a gateway over the past 6 million years. We now have results for understanding the outflow through the Gibraltar gateway, but the potential story that unfolds may be even more significant. The oceans and climate are inextricably linked. It seems there is an irrepressible signal of this nexus in contourite sediments.

We found a great deal more sand in the contourite sediments than anyone had expected, with a single, vast sand sheet spreading out nearly 100 km from the Gibraltar gateway, filling the channels as thick layers within the

67

contourites' mountains of mud. The sands testify to the great strength, high velocity and long duration of the Mediterranean bottom currents, and they could be ideal petroleum reservoirs if buried deeply enough in a trap. Typical petroleum reservoirs in such a setting are sands deposited by downslope turbidity currents, but these sands are different.

The leading scientific journal *Science* (see Chapter 13) published our expedition results in June 2014. A research article entitled 'Onset of Mediterranean Outflow into the North Atlantic' demonstrates that initial outflow into the Atlantic, following the opening of the Strait of Gibraltar, was relatively weak. Significant interaction between the outflow water and the North Atlantic did not begin until the late Pliocene, and that interaction affected global patterns of ocean currents and hence climate.

Climate-driven pulses of relatively warm salty water at depths of around 1,000 m over the last 3 million years contributed to increased global thermohaline circulation, which reduced pole-to-equator temperature gradients. These pulses coincided with widespread hiatuses in sediment deposition and pronounced changes in the sedimentary depositional patterns, and established the present-day sea floor morphology. The hiatuses and changes in depositional processes are related to regional tectonic events and margin instability, and appear to be related to overall plate reorganisation in the North Atlantic.

For me, this expedition was a new and exciting scientific experience. I was working with a team of 35 scientists from 14 countries. Working 12-hour shifts, seven days a week, we soon developed into a tight-knit team and colleagues soon became friends. While two months on a boat seems like a long time, there is always excitement and enthusiasm when new core arrives on deck, and we have the opportunity to see this sediment for the first time and to make new discoveries. There is also plenty of opportunity for social activities – this included a Christmas and New Year's party.

During my time onboard, not only was I able to utilise my existing skills as a sedimentologist, but also develop new skills. This was an excellent opportunity to interact with a wide variety of researchers including palaeoecologists, palaeomagnetists, geochemists and geophysicists.

For early-career researchers like me (my appearance is deceptive) this was an amazing opportunity, not only to develop new skills but also to be part of an experienced team of international researchers.

Figure 6.18. Craig Sloss working on cores on *JOIDES Resolution*
Source: Craig Sloss

I am continuing to work on the nature of contourites and deep marine sedimentation within the Gulf of Cadiz with shipboard colleagues. Specifically, I am using sediment characteristics to provide a detailed understanding of contourite deposition over the last 270,000 years. I aim to develop widely applicable models of how sediment supply, transport and bottom current flows are influenced by changes in climate and sea level. This was a fantastic opportunity for me to initiate research in a new area, not having worked on contourites before. Through this expedition and its research, I have had the opportunity to learn a lot more about deep marine sedimentation, specifically in relation to downslope and along slope processes. This has led to the development of a new research team at the Queensland University of Technology (QUT), which includes the involvement of two postdoctoral candidates, a PhD candidate and a Masters student, who now also have the opportunity to work on an international research project.

The experience I had on Expedition 339 has given me a drive and commitment to IODP research, which has led to being a shipboard scientist on Expedition 359 'Maldives Monsoon and Sea Level', and on Expedition 361 'Southern African Climates and Agulhas Current Density Profile' as a shore-based scientist with a research postdoc from QUT. I am excited about the opportunities I have with the Australian and New Zealand IODP Consortium (ANZIC) and IODP to further develop my research skills and to contribute to world-class international collaborations.

The *Resolution* Rookie off Costa Rica on IODP Expedition 344

Alan Baxter, then University of New England,
now McGill University, Canada

'So, Alan what are you up to for the next few months?'

'Actually, I'm spending two months on a ship, off the coast of Costa Rica, studying sediment core for 12 hours a day.'

'Really … why…?'

This was one of the most frequent exchanges I had with friends and family in the months preceding my first leg onboard the *JOIDES Resolution* (*JR*) as a member of IODP Expedition 344 science party. Why would anyone spend two months aboard ship, working a 12-hour shift every day, especially someone whose longest previous maritime experience was a ferry ride from Ireland to Wales (a four-hour journey)? The answer of course is what all earth science researchers would respond with: that a berth on the *JR* is something quite unique in the scientific world. It gives us the opportunity to conduct frontier science, in one of the world's most well-equipped labs, with world experts of different disciplines, working together on the same important scientific problems. A better question might have been: Why wouldn't anyone grab at that chance?

My adventure began in October 2012 when I left the University of New England in eastern Australia and flew for 26 hours through Sydney, Los Angeles and El Salvador to Panama, the point of departure for Expedition 344. Over the next few days, those who would become my shipmates arrived from all over the world. As the group began to get acquainted and talk to each other, two groups of scientists emerged, those who had been on previous expeditions and those, like myself, who were new to IODP. The old hands traded memories of expeditions past and jovially tried to scare the rookies with stories of sea sickness, huge waves and icy weather. Thankfully, we experienced none of this in calm equatorial waters off Costa Rica.

After hearing so much about the ship, and seeing numerous pictures of it online, my first glimpse of the *JR* was at the Panama Canal lock of Miraflores. Amongst the huge supertankers it looked quite small, but, with its derrick standing defiantly, the *JR* held its own. That afternoon we climbed aboard and the next two days were spent tied up at the dock, waiting to get the 'all clear' to depart. During that time, we were assigned a bunk and shown around the ship. I was surprised at how many modern conveniences and facilities were available, including a cinema, gym, games room and, most importantly, a great coffee machine. We were also introduced to the supporting crew, without whom we could not function, which included the IODP staff, sailors, drillers and canteen and cleaning staff.

While onboard we were also split into our discipline teams and shown our lab spaces. Luckily for our team, the palaeontology lab is one of the best on the ship and was great to work in. The micropalaeontology group was made up of foraminiferal experts from Spain, Japan and the US, a Costa Rican radiolarian specialist studying in Switzerland and myself, an Irishman who works in Australia, as the calcareous nannofossil biostratigrapher – truly an international consortium! Similar international compositions could be found in each discipline. As part of the micropalaeontology team, the job was to assign ages to the core as it came on deck. As most of the other groups had to wait up to four hours before the core was rested and then split, on occasion we had 'vultures' who lurked around the biostratigraphy lab to peck and peer at the sediment from the core catcher.

Our expedition was part of the Costa Rica Seismogenesis Project (CRISP), which was designed to elucidate the processes that control large earthquakes at erosional subduction zones (Harris et al., 2013). It is one of the few places in the world where the seismogenic zone is within reach of current drilling capabilities, and intersecting it is one of the long-term aims of the project.

I was on the night shift, which meant I started work at 12 am and finished at 12 pm. Although it is a long shift, it passed by quite quickly, especially when there was core on deck. The first few hours were spent catching up on work that had built up during the day, and then we would work on the samples taken during our shift. Working at night was fantastic; it gave me the opportunity to go out on deck and watch the diverse sea life surrounding the *JR* such as squid, turtles, dolphins, pilot whales, swordfish and tuna. In the morning, watching the sunrise behind the Central American Volcanic Arc was not a bad way to enjoy a coffee break!

Figure 6.19. *JOIDES Resolution* in the Panama Canal
Source: Alan Baxter

The most rewarding part of being on the ship was the integrated nature of the research and how supported one felt being in that working environment. Many questions that arose during discussions about the tectonics of the region were aided by the sedimentologists and elaborated on by geophysicists. Looking back, it was also great to be part of such a professional and seamlessly run setup, as any requests or problems were quickly fulfilled or solved. The technical staff onboard continually came to the rescue when dealing with rogue computer programs or updates to some of the data-recording software. This allowed the scientists to really focus on their work.

My time on the *JR* was one of the most stimulating and rewarding experiences I have had as a geologist – working towards the same goal with enthusiastic and talented colleagues from all over the world. I am still in contact with many of my shipmates and we have collaborated on many post-cruise studies.

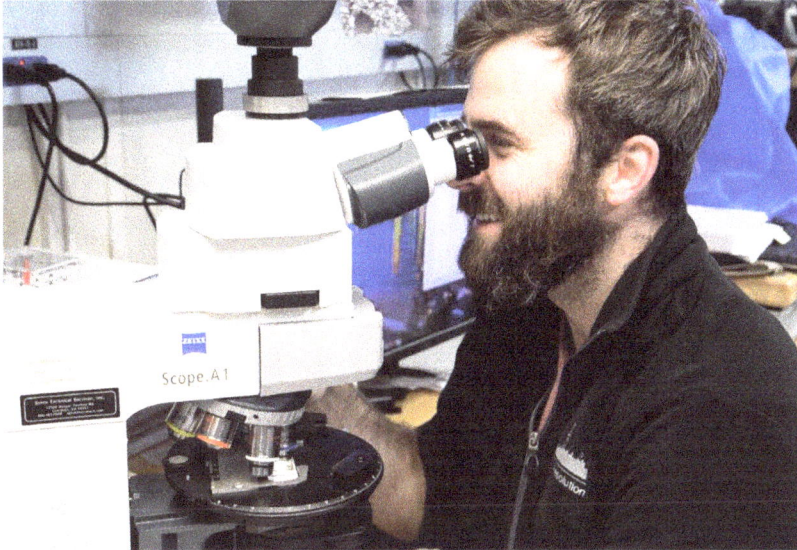

Figure 6.20. Alan Baxter examining sediment under a microscope aboard ship

Source: Alan Baxter

One of these studies (Schindlbeck et al., 2015) linked the source of ash layers observed at Site U1381 to Plinian eruptions that carried ash up to 450 km from the Galápagos hotspot. This is the first time that highly explosive Miocene volcanism has been reported from the Galápagos hotspot. In August 2016, I submitted a manuscript (Baxter et al., in review), in which we used new biostratigraphic, palaeomagnetic and tephrachronologic data to construct an age model for the incoming Cocos Plate, to a special volume of *Geochemistry, Geophysics, Geosystems* on 'Subduction processes in Central America with an emphasis on CRISP results'. Many journal articles and data reports have been published and more are still in the pipeline, including a synthesis paper summarising the CRISP expeditions. General preliminary results were outlined by Harris et al. (2013).

I have no doubt that I will apply again to be a part of another expedition. What will be fun is that I already have thought up a few stories to try and frighten the rookies.

References

Baxter, A.T., Kutterolf, S., Schindlbeck, J.C., Sandoval, M.I., Barckhausen, U., Li, Y.Z., and Petronotis K., In review. A CRISP Age Model for the Cocos Plate. In revision for the special theme on 'Subduction processes in Central America with an emphasis on CRISP results' in *Geochemistry, Geophysics, Geosystems*.

Harris, R.N., Sakaguchi, A., Petronotis, K., Baxter, A.T., Berg, R., Burkett, A., Charpentier, D., Choi, J., Diz Ferreiro, P., Hamahashi, M., Hashimoto, Y., Heydolph, K., Jovane, L., Kastner, M., Kurz, W., Kutterolf, S.O., Li, Y., Malinverno, A., Martin, K.M., Millan, C., Nascimento, D.B., Saito, S., Sandoval Gutierrez, M.I., Screaton, E.J., Smith-Duque, C.E., Solomon, E.A., Straub, S.M., Tanikawa, W., Torres, M.E., Uchimura, H., Vannucchi, P., Yamamoto, Y., Yan, Q., and Zhao, X., 2013. Expedition 344 summary. In Harris, R.N., Sakaguchi, A., Petronotis, K., and the Expedition 344 Scientists, *Proceedings of the IODP*, 344: College Station, TX (Integrated Ocean Drilling Program). doi.org/10.2204/iodp.proc.344.101.2013

Sandoval, M.I., Boltovskoy, D., Baxter, A.T., and Baumgartner, P.O., 2017. Neogene palaeoceanography of the eastern equatorial Pacific based on the radiolarian record of IODP drill sites off Costa Rica. *G3: Geochemistry, Geophysics, Geosystems*, 18(3): 889–906. doi.org/10.1002/2016GC006623

Schindlbeck, J.C., Kutterolf, S., Freundt, A., Straub, S., Wang, K., Jegen, M., Hemming, S.R., Baxter, A.T. and Sandoval, M.I., 2015, The Miocene Galápagos ash layer record of IODP Legs 334 & 344: Ocean-island explosive volcanism during plume-ridge interaction, *Geology*, 43(7): 599–602. doi.org/10.1130/G36645.1

A scientist off the southern Alaskan margin: IODP Expedition 341

Maureen Walczak (née Davies), then ANU, Canberra,
now Oregon State University, Corvallis

When I was a student, I devoted five years of graduate research to generating detailed palaeoenvironmental reconstructions of the northeast Pacific. My study was conducted using a suite of jumbo piston sediment cores collected in the Gulf of Alaska from aboard the RV *Ewing* in 2004; these cores had an average length of around 10–12 m and, in a few instances, they captured the most recent glacial termination ~15,000 years ago. I generated foraminiferal stable isotope, trace metal, faunal assemblage and radiocarbon data sets, as well as environmental and palaeomagnetic records, all of which were interpreted in the context of detailed CT scans of depositional structure and sedimentological history for a couple of these sites.

Figure 6.21. Maureen Davies (now Walczak) on a research cruise in the Arctic

Source: Paul Walczak

As convoluted as that series of analyses sounds, my PhD work was just a small part of a large interdisciplinary effort involving research institutions distributed throughout North America, with collaboration between my supervisors and the graduate students in our research group at Oregon State University, and the research groups of the other principal investigators involved in the 2004 *Ewing* expedition. Largely based on the success achieved with the sediments and seismic data of the *Ewing* research cruise, IODP elected to support what would eventually become Expedition 341: 'Climate and tectonics in the Gulf of Alaska'.

In mid-2011, I graduated and moved to a postdoctoral position at The Australian National University (ANU) supported by an Australian Research Council Super Science Fellowship to develop novel dating methods for high-latitude marine sediments. During the course of this appointment, I worked to develop new capacities for cosmogenic isotope analyses at ANU, focusing on techniques useful to glacial-proximal and carbonate-poor material. When the opportunity arose to apply for Expedition 341 through ANZIC, my supervisors very graciously supported my participation. Although I had worked for years on the survey materials for the proposed drill sites, many talented and very qualified scientists from around the world wanted to participate in the expedition and I knew that it was highly unlikely that I would be selected to join the science party. Indeed, in the first round of selections I failed to make the cut. However, when ANZIC was unexpectedly offered the opportunity to send a second scientist, AIO Program Scientist Neville Exon contacted me and encouraged me to resubmit my application. I was stunned when I was notified via email in December 2012 that I was to sail!

I was pretty thrilled when I arrived in Vancouver, Canada, to meet the *JOIDES Resolution* for the first time in my career. For me, this was the culmination of over five years of optimistic effort; however, the principal investigators and many of the original proponents of the Gulf of Alaska drill had been working towards this goal for close to two decades. Although limited space precluded the shipboard participation of every researcher involved in the site survey work, the first days of the cruise reunited me with numerous friends and colleagues. In addition to the familiar names and faces, capable researchers that I'd never met but would grow to admire arrived from Japan, China, Brazil, Germany, Spain, the United Kingdom, New Zealand and India. The science party was completed by two enthusiastic and talented teachers; one from New Zealand, who organised outreach to primary school students in many countries scattered over several continents.

Over the next two months, I learned a tremendous amount about the collection and shipboard processing of drill cores. While I'd been to sea many times before to collect gravity and piston cores, the material recovered by the IODP drill dwarfed anything I'd ever experienced. While working on the IODP cores didn't detract from the importance of the lower resolution and/or shorter timespan studies I'd participated in before, it certainly placed them in a humbling context. The lowest resolution of the jumbo piston cores that I worked on for my PhD captured the final throws of the last of the great Alaskan Pleistocene glaciations. On Expedition 341, we recovered expanded depositional sequences all the way back to the Miocene, capturing the birth of the first glaciers to develop on the Alaskan margin. The history captured in these records will be studied for decades and is already fundamentally changing our understanding of the interaction of ice, tectonics and the ocean in climatologically sensitive high-latitude environments.

There were some personal sacrifices associated with shipboard participation: even under mundane circumstances, two months is a long time to be away from loved ones, and sadly my grandmother passed away unexpectedly while we were at sea. My then fiancé and I were also planning our wedding for that summer before I received my invitation to sail, and our hastily rescheduled ceremony was only a week after the *JOIDES Resolution* disgorged the science party in Valdez, Alaska. However, my participation in Expedition 341 has been the most productive ongoing experience I've ever had, both scientifically and professionally. In addition to the personal educational aspects of participation, I formed new collaborations with other dynamic young scientists from around the world and strengthened established collaborations with my earlier mentors. I have already co-authored several publications on shipboard and early shore-based results from this expedition, data generated from expedition samples in the last 18 months of my postdoc at ANU form the basis of several more papers in progress, and additional preliminary results are the fodder for research proposals that will likely drive my career for the next several years.

To other early-career scientists considering applying for shipboard, or even shore-based, participation in an IODP Expedition: it's a life-changing experience and I wish you the very best of luck. Don't be discouraged if you don't get on the first expedition you apply for (IODP Expedition 341 was my third attempt to sail over several years and my first success). It really is well worth all the patience and effort.

Two months off Alaska as an educator: IODP Expedition 341

Carol Larson, then Education Officer, now Education Team Leader, National Aquarium of New Zealand, Napier

You would have thought I'd won the lottery when I was accepted to spend two months at sea on the IODP Expedition 341 'Southern Alaska Margin' as an education officer!

I was especially excited to be aboard a ship again as I had spent two marvellous months on stern trawlers in the Bering Sea as a fisheries biologist/observer during my university days. And yet, I was a bit daunted by the fact that I was just a teacher/biologist on a geological research vessel with an international team of high-level scientists. My fears were allayed as soon as I boarded the *JOIDES Resolution* and met the other education officer, Alison Mote from Texas, and the rest of my 123 shipmates: scientists, engineers, operational and galley crew. Everyone was very welcoming and friendly. Many of the team had done this before and they were dab hands at assisting newcomers.

From the outset, I was very pleased to see that the science team was half female. This was something I could tell our Careers in Science students at the National Aquarium of New Zealand when I got back home!

Figure 6.22. Carol Larson on the captain's bridge

Source: Alison Mote, Education Officer from Austin, Texas, who was working with Carol Larson

All aboard were wonderful people who'd led very interesting lives all over the globe, so there was plenty to talk about. There was never a dull moment, and to keep morale up (or raise it even higher) we had parties for birthdays and holidays and played the odd joke on each other. We would decorate the labs with funny pictures and put googly eyes in unusual places. We made awards and certificates for the scientists for their various achievements. We did a pet wall, matching up whose pet belonged to whom. We also arranged 'ice-cream surprises', where we would cheerfully deliver ice-cream to the working teams.

The stewards and crew looked after us extremely well. Everyone worked in shifts. We shared spaces and took turns at eating and sleeping. The living quarters were very comfortable yet snug – I soon understood the phrase 'shipshape'. Our rooms were cleaned and laundry done as often as needed. We didn't even need to make our beds! The food was top-class and plentiful. I walked the deck and exercised at the gym to keep off the kilos.

Alison and I roomed and worked together on education outreach to both hemispheres seven days a week (6 am to 6 pm). We created impact by video-conferencing (alongside our onboard scientists) with over 2,000 participants from 54 schools, museums, summer camps and aquariums in Australia, NZ, UK and the US. We posted daily on Facebook, Twitter and wrote a blog on the *JOIDES Resolution* website. In true all-hands-on-deck fashion, we helped the scientists with other jobs like sieving mud for microfossils, using the SEM camera to photograph fossils and preparing slides. This gave us even more insight as educators and, when we could, we roamed the ship looking for novel and interesting blog posts and exciting photographs for creating competitions on Facebook. We were allowed to tour and access all parts of the ship, including the captain's bridge. All the amazing navigational technology, the rumbling engine room, even the jam-packed kitchen stores and giant freezers full of food were fascinating – and good educational fodder.

As a biologist, it was exciting to see northern fur seals, sei whales, humpback whales, orca, black-footed albatross, short-tailed shearwaters and many other seabirds. The grey Alaskan skies and the ever-changing sea were a daily art gallery – always open – to observe and revel in.

In the two years since I participated on the cruise, I have organised and run a geology camp for Year 7–10 students at the National Aquarium with the help of the fantastic GNS Science Education Team and with sponsorship from them, the New Zealand Royal Society, the Hawkes Bay Regional Council and TAG Oil. I have given talks on my Alaskan experience to the Hawkes Bay Branch of the Royal Society, Rotary and many other community and school groups. I have also become involved with the East Coast LAB group, which seeks to educate coastal communities from the East Cape to Wellington on earthquake and tsunami preparedness.

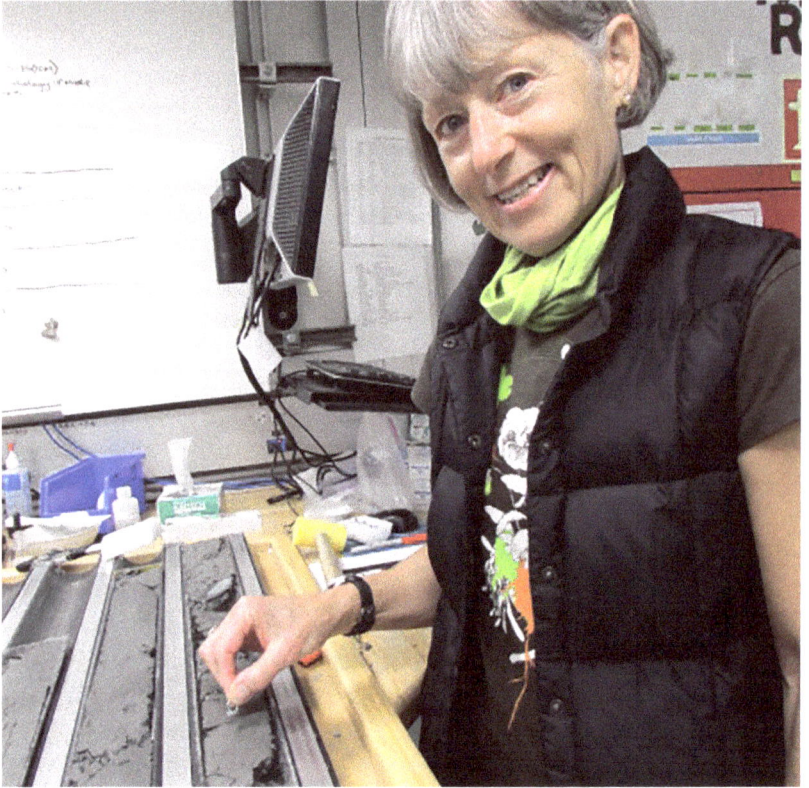

Figure 6.23. Carol Larson with core samples on the sampling table
Source: Alison Mote, education officer from Austin, Texas

The experiences I had and the contacts I made on the *JOIDES Resolution* have given me the resources and confidence to branch out into this new area of science. It was a wonderfully rewarding and fulfilling experience.

Would I go again? Most definitely, but maybe to the tropics next time!

Investigating deep plutonic crust in Hess Deep: IODP Expedition 345

Trevor Falloon, University of Tasmania, Hobart

'Could Trevor Falloon please report to the Mess' – the dreaded announcement was broadcast loudly across the entire ship. Like a man condemned, I reluctantly made my way there. Waiting for me was a small gathering of the scientific party with smiles and smirks on their faces – I tried to look pleased and excited – eventually the cooks came out with a cake and everybody started singing 'Happy Birthday!' This is one of the many dangers and difficulties to be faced when you volunteer to go on an IODP drilling expedition for two months or so.

I boarded the *JOIDES Resolution* on 13 December 2012 at Puntarenas, Costa Rica. It was the beginning of Expedition 345 to Hess Deep in the eastern equatorial Pacific, well west of Colombia. Boarding was not the leisurely stroll up a gangway that I had expected. The ship was out in the harbour and we had to take a small boat and then board by a rope ladder. It looked easy enough, but I wasn't so happy once on the ladder, trying to pull myself up and maintain grip and balance – lest I fall into the murky depths below! Once onboard, we all had to surrender our passports to the ship's captain. This was normal procedure and I had done this many times before on other ships. However, I wasn't expecting that the passport information would be passed on to the publication officer who quickly made a list of all those poor souls who had birthdays during our time at sea. Alas, this included me and hence my obligatory acknowledgement of the birthday cake and one's mortality.

Another difficulty with this particular expedition was that our time at sea would also include Christmas Day. I had been in this situation before – 20 years ago! Then I was a participant on ODP Leg 147 also to Hess Deep in 1992–1993. In those days, Christmas celebrations onboard were not organised in any way – things were sort of worked out on the spot at sea. I can still remember the small Christmas concert we had in the science lounge – with one of the scientists in full Texas cowboy outfit singing and playing his guitar, followed by a very large science technician performing, of all things, a belly dance! That really put me off my Christmas lunch. The chef's bizarre *piece de resistance* was a huge Christmas tree made out of cooked prawns, which nobody had the appetite to eat.

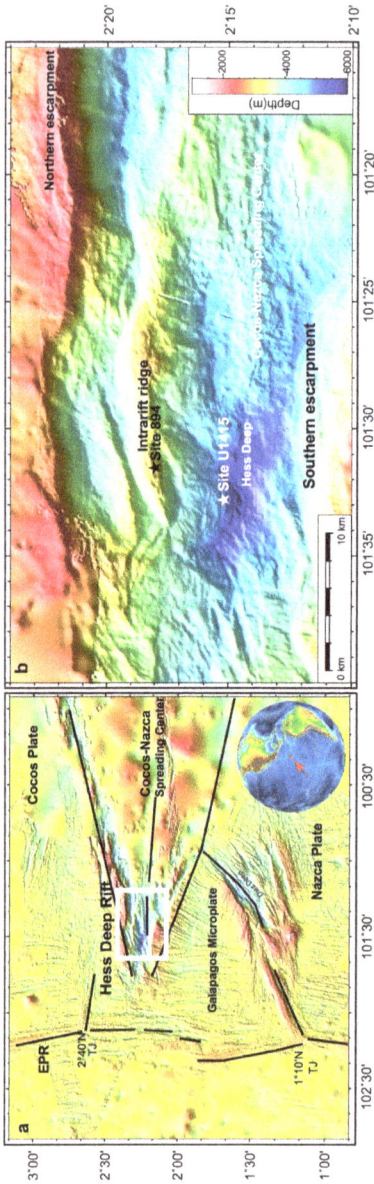

Figure 6.24. Maps of study area
(north to the left of page)

Source: Gillis et al. (2014). Primitive Layered
Gabbros from Fast-Spreading Lower
Oceanic Crust. *Nature*, 505(7482): 204–207

Not so with Expedition 345. We all had to bring a small gift to give to another member of the science party, and we were asked in advance to prepare an item for the Christmas concert. So when I boarded I had a small calendar of Australian scenes (how original!) and a pocket trumpet. I had bought the trumpet off the internet from some dodgy site especially for the voyage. You get what you pay for. It was very difficult to play and keep in tune; some key combinations were so far out that I had to avoid playing those notes. I don't mean to blow my own trumpet, but it was great to have an instrument with me, as I was able to help out by accompanying the carol singing (which occurred throughout all the ship's stations over several days) and perform at the Christmas concert. The concert was a hoot and everybody had a brilliant time.

Other difficulties were the scientific drilling itself. The aim of Expedition 345 was to drill into the lower oceanic crust at Hess Deep to test current models of oceanic crustal formation and cooling among other things. As with Leg 147, the drilling was fraught with technical difficulties and very low rock recovery. However, we made up in quality what we lacked in quantity, managing to recover spectacular examples of layered gabbros and primitive lower crustal rocks. The shipboard results themselves were enough to have a paper published in *Nature* (see Chapter 13). One problem that was very difficult to solve and caused a great deal of anxiety and tension within the science party was how to fairly sample the core when the recovery was so limited. Not everybody could have a piece of some of the more prized rock types. The chief scientists did a fantastic job in coordinating the science party into collaborative research teams for sampling, and all went well in the end.

Figure 6.25. Trevor Falloon warming up before the Christmas concert performance
Source: Shipboard scientific party, Expedition 345

One of the great things about ocean drilling expeditions is the opportunity to interact with an international cast of scientists, all dedicated to the same scientific goals and all working hard to do the best job they can whilst onboard. These experiences lead to lifelong friendships and collaborations. As a result of my involvement on Expedition 345, I spent two months in Toulouse working on the oceanic gabbros – a great experience. I have no hesitation in recommending involvement in an IODP expedition for any Australasian scientist – young or old, it is a very rewarding experience.

7

A sampler of scientific results

These articles give an idea of the variety of drilling expeditions in our region in the period 2010 to 2013. They outline early results from these scientifically very diverse expeditions and were written in 2015 and 2016.

》

Sediments, rocks and chemical fossils in the Canterbury Basin, New Zealand: IODP Expedition 317

Simon C. George, Macquarie University, Sophia Aharonovich, Macquarie University, Greg H. Brown, GNS Science, and Julius S. Lipp, University of Bremen

Introduction

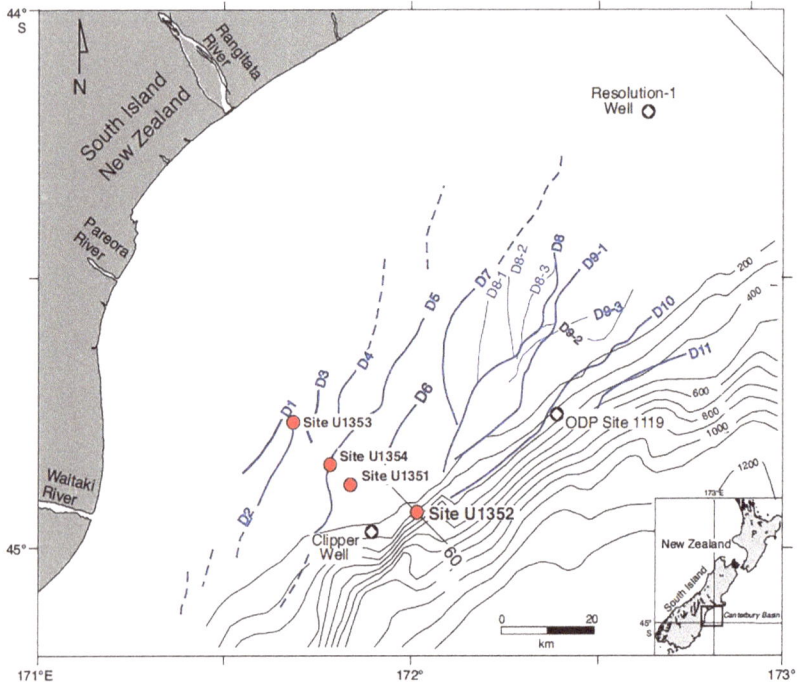

Figure 7.1. Drilled and proposed Expedition 317 sites, together with seismic grids, exploration wells Clipper and Resolution, and Ocean Drilling Program Site 1119

Source: Modified from Expedition 317 Scientists (2011)

IODP Expedition 317 drilled the Canterbury Basin east of the South Island of New Zealand and was devoted to understanding the relative importance of global sea level (eustasy) versus local tectonic and sedimentary processes in controlling continental margin sedimentary cycles (Expedition 317 Scientists, 2011). Sediments were recovered from four sites (Figure 7.1), including three on the continental shelf (landward to basinward, Sites U1353, U1354, U1351), and one on the continental slope (Site U1352). The drilled sediments range in age from Late Eocene to Holocene, and provide a stratigraphic record of depositional cycles across a shallow marine transect most directly affected by relative sea-level change. Sedimentation is thought to have been controlled by the timing of uplift of the adjacent Southern Alps, as well as the influence of strong ocean currents, including the Antarctic Circumpolar Current and the Deep Western Boundary Current. The sedimentary record is being used to estimate the timing and amplitude of global sea-level change, and to document the sedimentary processes that operate during sequence formation. Sites U1353 and U1354 provided significant, double-cored, high-recovery sections through the Holocene and late Quaternary for high-resolution study of recent glacial cycles in a continental shelf setting (Expedition 317 Scientists, 2011).

The transformation of sediment into rock

At Site U1352, a continuously cored 1,927 m thick Holocene to Late Eocene section uniquely documents downhole changes in induration, from unlithified sediments to rock, using a wide range of petrological, petrophysical and geochemical data sets (Figure 7.2). Porosity decreases from around 50 per cent at the surface to about 10 per cent at the base of the hole, with a corresponding increase in density from around 2 to 2.5 g cm^3 (Marsaglia et al., 2017). There are progressive changes in the minerals with depth, including an increase in carbonate and a decrease in quartz and clay content. Grain compaction is first seen in rock thin sections at 347 m below sea floor (mbsf). Pressure solution begins at 380 m and is common below 1,440 m, with stylolite development below 1,600 m, and sediment injection features below 1,680 m. Pore water geochemistry (Figure 7.2) and petrographic observations document two active zones of cementation, one shallow (down to ~50 m), driven by methane oxidation by sulphate, transitioning to another burial-related cementation zone starting at ~300 m, resulting from carbonate dissolution and re-precipitation. Carbonate cementation becomes more common with depth. These results quantify downhole diagenetic changes and verify depth estimates for these processes inferred from outcrop and other well-based studies.

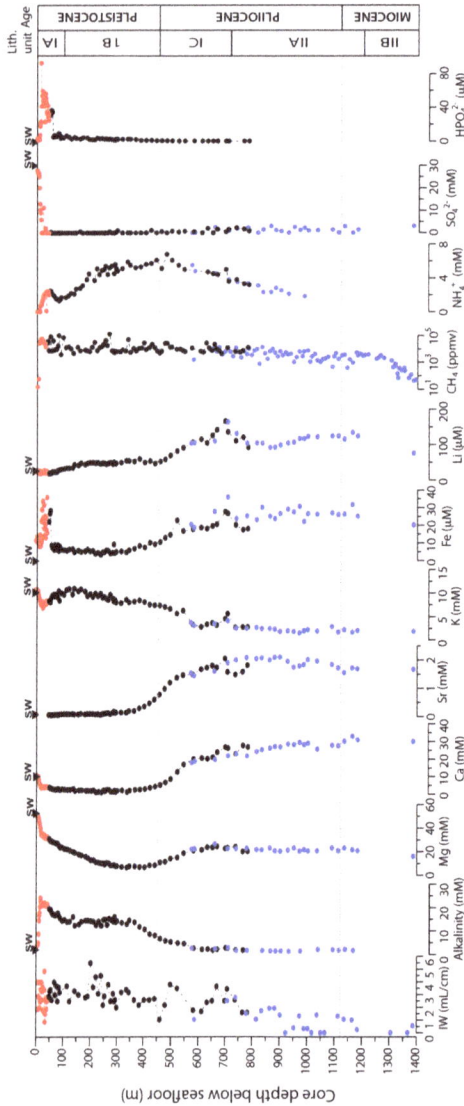

Figure 7.2. Downhole variation of methane from headspace analysis, and interstitial water, alkalinity and geochemical parameters relative to lithostratigraphy in Holes U1352A (red symbols), U1352B (black symbols) and U1352C (blue symbols)

Lithostratigraphic units I to II and their subdivisions are indicated. Parameters are methane (ppmv), yield of interstitial water (IW) per cm of squeezed whole-round cores; alkalinity (mM); sulphate (mM); magnesium (mM); calcium (mM); strontium (mM); potassium (mM); iron (µM); lithium (µM); ammonium (mM); and phosphate (µM). IAPSO seawater values are shown with the black-filled triangle and marked with 'SW'.

Source: From Marsaglia et al. (2017)

Chemical fossils to define source input and palaeoceanography

Coastal ecosystems such as the Canterbury Basin are characterised by high biological productivity due to a significant amount of organic matter (OM) arriving from oceanic or terrigenous inputs. Local tectonic activity in New Zealand, such as the uplift of the Southern Alps and volcanic eruptions, created a significant input of eroded sediment and terrigenous OM into the marine environment. Marine productivity, on the other hand, was influenced by palaeoceanographic changes such as global sea-level variations, changes in currents and water temperature variation. The OM from the continental margin area contains information about both types of organic input. Hydrocarbons including chemical fossils (biomarkers) preserved in cored sediments from IODP Expedition 317 have been used to define the source of the OM and to determine sea surface palaeo-temperature reconstructions of the area. Rock-Eval pyrolysis (Expedition 317 Scientists, 2011) results combined with the distribution of aromatic hydrocarbons and biomarkers show good preservation of the OM (Figure 7.3). Calculated vitrinite reflectance based on the latter (Peters et al., 2005) varies from the early oil window in the Oligocene to immature for the Late Miocene and younger sediments.

The Canterbury Basin sediments contain high amounts of marine OM during the Oligocene and Early Miocene epochs (Figure 7.3), as shown by n-alkane profiles with little odd-over-even carbon number predominance, as measured by the low carbon preference index (Bray and Evans, 1961) and the low amount of aquatic plants (Paq = Proxy ratio that distinguishes between terrestrial plants and aquatic plants using mid-chain length n-alkanes; Ficken et al., 2000). The Middle Miocene to Pliocene section is characterised by an upward increase in terrigenous OM input, with carbon preference index values increasing to much greater than one, likely due to greater input from terrigenous vascular plants as the Southern Alps were uplifted (Figure 7.3). Similar trends in OM input can also be recognised based on the total organic carbon (TOC) and total nitrogen (TN) data collected aboard (Expedition 317 Scientists, 2011). Relatively low TOCdiff/TN ratios of up to 12 in the Early Miocene indicate predominant input of marine OM. Increasing TOCdiff/TN values above 20 through the Middle and Late Miocene suggest a mix of marine and terrigenous OM input, with greater influence of land-derived vegetation (Meyers, 1997).

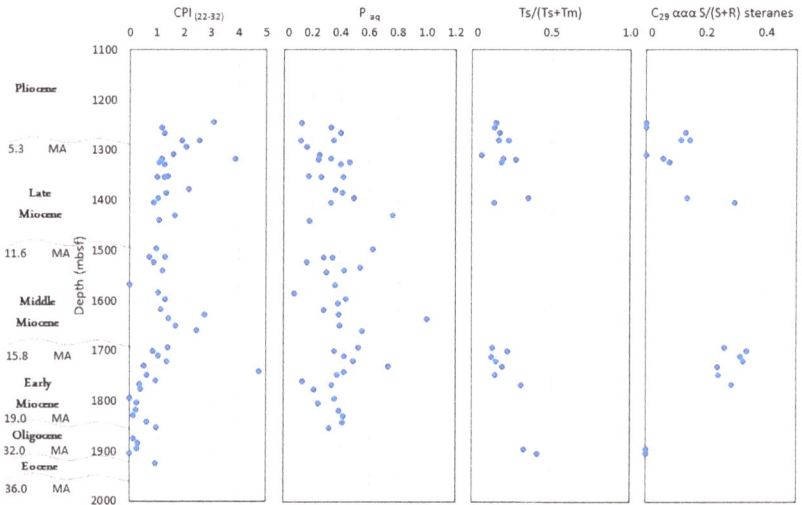

Figure 7.3. Biomarker and bulk geochemistry data from the continental slope site (U1352)

The ratios are: carbon preference index = (2*C23+C25+ C27+C29+C31)/(C22+C24+ C26+C28+C30+C32); Paq = (C23+C25)/(C23+C25+C29+C31). The biomarkers are: Ts = C27 18α-trisnorneohopane; Tm = C27 17α-trisnorhopane; C27 and C29 5α,14α,17α(H) 20R steranes. Total organic carbon measured with source rock analyser (TOCSRA), total organic carbon vs total nitrogen (TOCdiff/TN). TOC was measured aboard ship by calculating the difference between total carbon and carbonate carbon.

Source: Expedition 317 Scientists (2011)

The distribution of archaeal membrane lipids such as glycerol dialkyl glycerol tetraethers is being used to reconstruct palaeo sea-surface temperatures (SST), using the tetraether index of 86 carbon atoms (TEX86) (e.g. Schouten et al., 2002; Kim et al., 2010; Schouten et al., 2013). In addition, it has been suggested that the branched glycerol dialkyl glycerol tetraethers are derived predominantly from a terrestrial environment (e.g. Hopmans et al., 2004; Schouten et al., 2013). Varying input of terrigenous sediment into a marine environment can be calculated by the branched and isoprenoid tetraether index, and this generally shows a decrease of terrigenous input with increasing distance from deltas and river fans (e.g. Weijers et al., 2006; Kim et al., 2012). The Canterbury Basin sediments have been used to reconstruct palaeo SST. Initial data collected by Dr Julius Lipp were presented during the post-cruise workshop in 2011 in New Zealand and show decreasing temperatures from the Oligocene to the Holocene. In 2015, new collaboration between Sophia Aharonovich and Dr Lipp led to the sampling resolution being increased by adding an additional data set from Oligocene and Miocene samples on the continental slope site U1352 (Figure 7.4). The results show a convincing trend of decreasing palaeo SST from around 30°C during the Late Palaeocene period to around 15°C during the Late Miocene. The development of the Antarctic Circumpolar Current and general changes in the seaways during the latter part of the Cenozoic may play a significant role in controlling the observed SST change in the eastern part of New Zealand (Lawver and Gahagan, 2003; Barker and Thomas, 2004).

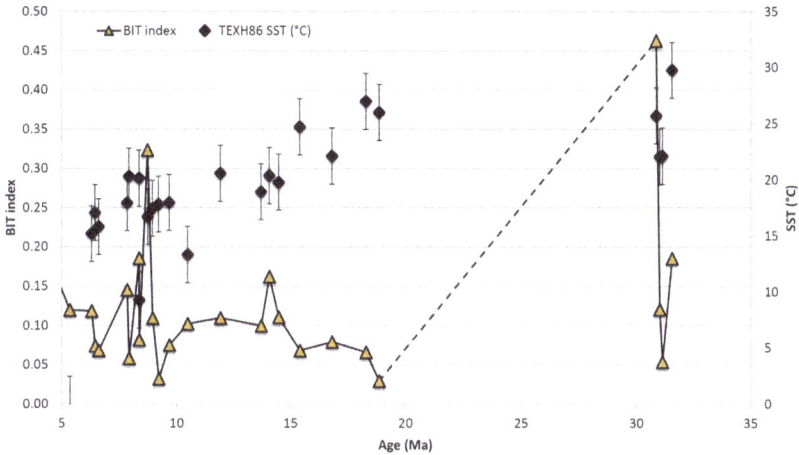

Figure 7.4. Reconstructed sea-surface temperatures (SST) using the TEX86 index from the continental slope site (U1352)

The calculated branched and isoprenoid tetraether index (Weijers et al., 2006) shows variations in terrestrial input into the marine sediments.

Source: After Kim et al. (2010); and authors' unpublished data

Conclusions

Our work and that of the other members of the scientific team shows that sedimentation was controlled by the timing of uplift of the adjacent Southern Alps, as well as the influence of strong ocean currents, including the Antarctic Circumpolar Current and the Deep Western Boundary Current. The sedimentary record also allows us to estimate the timing and amplitude of global sea-level change.

References

Barker, P.F., and Thomas, E., 2004. Origin, signature and palaeoclimatic influence of the Antarctic Circumpolar Current. *Earth-Science Reviews* 66: 143–162. doi.org/10.1016/j.earscirev.2003.10.003

Bray, E.E., and Evans, E.D., 1961. Distribution of n-paraffins as a clue to recognition of source beds. *Geochimica et Cosmochimica Acta* 22(1): 2–15.

Expedition 317 Scientists, 2011. Expedition 317 Summary, in Fulthorpe, C.S., Hoyanagi, K., Blum, P., Expedition 317 Scientists (eds), *Proceedings of the IODP*, 317: Tokyo (Integrated Ocean Drilling Program Management International, Inc.).

Ficken, K.J., Li, B., Swain, D.L., and Eglinton, G., 2000. An n-alkane proxy for the sedimentary input of submerged/floating freshwater aquatic macrophytes. *Organic Geochemistry* 31: 745–749. doi.org/10.1016/S0146-6380(00)00081-4

Hopmans, E.C., Weijers, J.W.H., Schefuß, E., Herfort, L., Sinninghe Damsté, J.S., and Schouten, S., 2004. A novel proxy for terrestrial organic matter in sediments based on branched and isoprenoid tetraether lipids. *Earth and Planetary Science Letters* 224: 107–116. doi.org/10.1016/j.epsl.2004.05.012

Kim, J.H., Van der Meer, J., Schouten, S., Helmke, P., Willmott, V., Sangiorgi, F., Koc, N., Hopmans, H.C., and Sinninghe Damsté, J.S., 2010. New indices and calibrations derived from the distribution of crenarchaeal isoprenoid tetraether lipids: Implications for past sea surface temperature reconstructions. *Geochimica et Cosmochimica Acta* 74: 4639–4654. doi.org/10.1016/j.gca.2010.05.027

Kim, J.-H., Zell, C., Moreira-Turcq, P., Pérez, M.A.P., Abril, G., Mortillaro, J.-M., Weijers, J.W.H., Meziane, T., and Sinninghe Damsté, J.S., 2012. Tracing soil organic carbon in the lower Amazon River and its tributaries using GDGT distributions and bulk organic matter properties. *Geochimica et Cosmochimica Acta* 90: 163–180.

Lawver, L.A., and Gahagan, L.M., 2003. Evolution of Cenozoic seaways in the circum-Antarctic region. *Palaeogeography, Palaeoclimatology, Palaeoecology* 198: 11–37.

Marsaglia, K.M., Browne, G.H., George, S.C., Kemp, D.B., Jaeger, J.M., Carson, D., Richaud, M. and IODP Expedition 317 Scientific Party, 2017. The transformation of sediment into rock: insights from IODP Site U1352, Canterbury Basin, New Zealand. *Journal of Sedimentary Research* 87: 272–287.

Meyers, P.A., 1997. Organic geochemical proxies of palaeoceanographic, palaeolimnologic, and palaeoclimatic processes. *Organic Geochemistry* 27: 213–250.

Peters, K.E., Walters, C.C., and Moldowan, J.M., 2005. *The Biomarker Guide*, 2nd Edition. Press Syndicate of the University of Cambridge.

Schouten, S., Hopmans, E.C., and Damste, J.S.S., 2013. The organic geochemistry of glycerol dialkyl glycerol tetraether lipids: A review. *Organic Geochemistry* 54: 19–61. doi.org/10.1016/j.orggeochem.2012.09.006

Schouten, S., Hopmans, E.C., Schefuß, E., and Sinninghe Damsté, J.S., 2002. Distributional variations in marine crenarchaeotal membrane lipids: A new tool for reconstructing ancient sea water temperatures? *Earth and Planetary Science Letters* 204: 265–274. doi.org/10.1016/S0012-821X(02)00979-2

Weijers, J.W.H., Schouten, S., Spaargaren, O.C., and Sinninghe Damsté, J.S., 2006. Occurrence and distribution of tetraether membrane lipids in soils: implications for the use of the TEX 86 proxy and the BIT index. *Organic Geochemistry* 37: 1680–1693. doi.org/10.1016/j.orggeochem.2006.07.018

Wilkes Land climatic and oceanographic changes: IODP Expedition 318

Rob McKay, Victoria University of Wellington, and Kevin Welsh, University of Queensland

The East Antarctic Ice Sheet is the world's largest ice sheet, and would be the equivalent to 60 m of sea-level rise if it were all to melt. The history and evolution of the ice sheet is likely to have dominated the Earth's climate during the geological past, affecting global sea levels, planetary albedo, oceanic overturning and surface currents, the productivity of the Southern Ocean and, by extension, atmospheric CO_2 content. Despite this, there are remarkably few records that are capable of reconstructing its evolution and dynamics throughout its geological history.

Expedition 318 to Wilkes Land in East Antarctica in 2010 was the first visit by the Integrated Ocean Drilling Program (or its predecessor program) to the Antarctic for a decade. The primary aim was to understand the evolution and dynamics of this ice sheet since its inferred inception at the Eocene–Oligocene boundary. This region of Antarctica is of interest because (unlike other sectors of the East Antarctic Ice Sheet that are grounded above sea level) the ice sheet here sits on the Earth's surface up to 2,000 m below sea level (mbsl), making it potentially sensitive to marine instability processes – in particular, glacial melting resulting from moderate changes in ocean temperature and warm waters upwelling next to the ice sheet. To capture these phenomena, a transect of drill cores was collected from shallow continental shelf drill sites to deep-water continental rise/abyssal plains to investigate oceanographic linkages to changes in continental climate and ice sheets in East Antarctica. The principal goals of the expeditions were to:

- establish the timing and nature of the first arrival of ice at the Wilkes Land margin, inferred to have occurred during the earliest Oligocene (~34 million years ago (Ma))
- reconstruct variations in the volume of the East Antarctic Ice Sheet and oceanographic/biological changes during past climatic warm events and transitional climate states
- obtain an ultra high-resolution (i.e. annual) Holocene sediment record of climate and oceanographic variability.

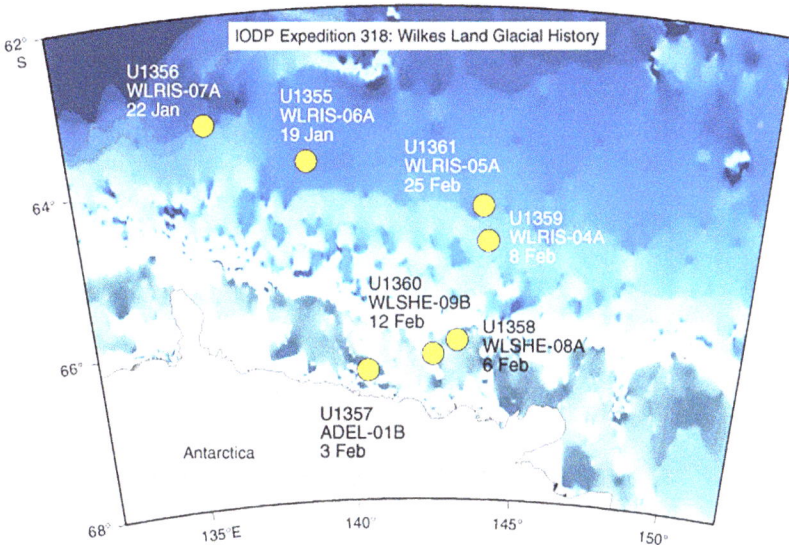

Figure 7.5. Map showing Expedition 318 drill sites off Wilkes Land
Source: Escutia et al. (2011)

A total of ~2,000 m of high-quality middle Eocene–Holocene sediments were recovered from water depths of between ~400 and 4,000 mbsl, and together the cores represent ~55 million years of Antarctic history. These cores provide an unprecedented history of the transition from an ice-free 'greenhouse Antarctica' and the cooling and onset of continental-scale glaciation of the East Antarctic Ice Sheet (Escutia et al., 2011).

Fossilised pollen enabled a reconstruction of 'Greenhouse Antarctic climates' during the early Eocene (between 55 and 48 Ma). These data indicate that coastal regions of Wilkes Land were characterised by a lowland, warm rainforest dominated by tree ferns, palms and trees belonging to the Bombacaceae family whose modern relatives are found on Madagascar (Pross et al., 2012). Superimposed on this coastal pollen assemblage was an upland, mountain forest region with beech trees and conifers, revealing a more temperate climate in the East Antarctic interior and highlands. The pollen indicates that temperatures on the Antarctic coast were on average around 16°C, with summers reaching 21°C. Importantly, winters were warmer than 10°C despite Antarctica being in nearly the same position it currently is, with 24-hour darkness during winter months. Organic molecules preserved from Eocene soil bacteria confirm that the temperature was at least as warm as the pollen indicates.

Following this early Eocene warmth, a period of mid-Eocene cooling was also reconstructed from the analyses on fossil algae (dinoflagellate cysts) and organic biomarker proxies. These show cooling surface waters and Antarctic air temperatures coeval with the development of oceanic circulation through the Tasmanian Gateway. It is hypothesised that the onset of the westbound Antarctic Counter Current terminated the early Eocene hothouse, and continued cooling ultimately gave rise to the development of continental-scale glaciations on Antarctica (Bijl et al., 2013).

A continuous sedimentary record of the exact transition from the Eocene 'greenhouse' to the Oligocene 'icehouse' still remains elusive, as it is represented by a major hiatus in the IODP Expedition 318 cores – a common theme in previous Antarctic drill cores. However, earliest Oligocene (~34 Ma) glacially influenced sediments recovered from the continental shelf Site U1360 indicate that by this time a continental-scale East Antarctic Ice Sheet had extended to near the continental shelf edge (Escutia et al., 2011). This was accompanied by large relative sea-level variations, which probably exceed eustatic sea-level fall, causing local deepening (Stocchi et al., 2013). Fossil marine dinoflagellate cyst records indicate that the glacial onset and sea ice were associated with a fundamental regime shift in zooplankton–phytoplankton interactions and community structure in the Southern Ocean by the earliest Oligocene (Houben et al., 2013).

One of the longest running debates in Antarctic geosciences concerns the relative stability of the marine sectors of the East Antarctic Ice Sheet since its inception. Of particular importance is the Pliocene epoch, when global temperatures were 2–3°C higher than today and atmospheric CO_2 was ~400 ppm (parts per million), thus providing an insight into ice sheet response to climates similar to those predicted for the next century as a consequence of anthropogenic climate change. Marine mud deposited offshore of East Antarctica during the Pliocene revealed a geochemical fingerprint that enabled the science team to trace where it came from on the continent (Cook et al., 2013). They discovered that the mud originated from rocks that are currently hidden under the ice sheet, and would only be eroded by an ice sheet that had retreated inland. The scale of this retreat, combined with the loss of the smaller West Antarctic and Greenland Ice Sheets may have resulted in sea level rises of ~20 m above present-day levels. A study by Patterson et al. (2014) identified pulses of iceberg discharge from the East Antarctic associated with these major ice sheet retreat events between 4 and 2 million years ago. They showed there was a major shift in the timing and intensity of iceberg discharge between 3.5 and 2.5 million years ago. Between 3.5 and 2.5 million years ago, naturally declining CO_2 levels (to 280 ppm) resulted in climate cooling and expansion of Southern Ocean sea ice. This sea ice cover prevented wind-driven warm water currents from penetrating far enough south to melt the ice sheets, and thus iceberg discharge decreased significantly at this time, indicating that the East Antarctic Ice Sheet was in general less likely to collapse when it was protected by its fringing sea ice belt.

Work continues on many other studies from cores collected during this cruise. Of particular interest are the first results to be published from Site U1357, a ~180 m long sediment core with annual-scale laminae present through the entire Holocene. This is a unique sediment core situated in one of the three main regions of Antarctic Bottom Water formation that has an ice-core or tree-ring style record. This core has the potential to fundamentally alter our understanding of natural variability of physical, chemical and biological processes in the high polar latitudes.

Figure 7.6. Icebergs are spawned from the Antarctic continent in the background

Source: Rob Dunbar, Stanford University

Figure 7.7. A school video broadcast from the ship with scientists talking about a core

Source: Rob Dunbar, Stanford University

Figure 7.8. Rob McKay analysing an Adelie Drift core collected offshore of East Antarctica

This core contains near annual layers of diatoms (marine algae) and is the most expanded Holocene sediment sequence ever recovered from the world's oceans.

Source: Rob Dunbar, Stanford University

References

Bijl, P.K., Bendle, J.A.P., Bohtay, S.M., Pross, J., Schouten, S., Tauxe, L., Stickley, C.E., McKay, R.M., Rohl, U., Olney, M., Sluijs, A., Escutia, C., Brinkhius, H., Welsh, K., and the IODP Expedition 318 Scientists, 2013. Eocene cooling linked to early flow across the Tasmanian Gateway. *Proceedings of the National Academy of Sciences of the United States of America*, 110 (24): 9645–9650. doi.org/10.1073/pnas.1220872110

Cook C. P., Van De Flierdt T., Williams T., Hemming S. R., Iwai M., Kobayashi M., Jimenez-Espejo F. J., Escutia C., Gonzalez J. J., Khim B.-K., McKay R. M., Passchier S., Bohaty S. M., Riesselman C. R., Tauxe L., Sugisaki S., Galindo A. L., Patterson M. O., Sangiorgi F., Pierce E. L., Brinkhuis H., Klaus A., Fehr A., Bendle J. a. P., Bijl P. K., Carr S. A., Dunbar R. B., Flores J. A., Hayden T. G., Katsuki K., Kong G. S., Nakai M., Olney M. P., Pekar S. F., Pross J., Rohl U., Sakai T., Shrivastava P. K., Stickley C. E., Tuo S., Welsh K., and Yamane M., 2013. Dynamic behaviour of the East Antarctic ice sheet during Pliocene warmth. *Nature Geoscience,* 6: 765–769. doi.org/10.1038/ngeo1889

Escutia, C., Brinkhuis, H., Klaus, A., Fehr, A., Williams, T., Bendle, J.A.P., Bijl, P.K., Bohaty, S.M., Carr, S.A., Dunbar, R.B., Gonzàlez, J.J., Hayden, T.G., Iwai, M., Jimenez-Espejo, F.J., Katsuki, K., Kong, G.S., McKay, R.M., Nakai, M., Olney, M.P., Passchier, S., Pekar, S.F., Pross, J., Riesselman, C., Röhl, U., Sakai, T., Shrivastava, P.K., Stickley, C.E., Sugisaki, S., Tauxe, L., Tuo, S., van de Flierdt, T., Welsh, K., and Yamane, M., 2011. *Proceedings of the IODP,* 318: Tokyo. (Integrated Ocean Drilling Program Management International, Inc.).

Houben, A.J.P., Bijl, P.K., Pross, J., Bohaty, S.M., Passchier, S., Stickley, C.E., Röhl, U., Sugisaki, S., Tauxe, L., van de Flierdt, T., Olney, M., Sangiorgi, F., Sluijs, A., Escutia, C., Brinkhuis, H., Klaus, A., Fehr, A., Williams, T., Bendle, J.A.P., Carr, S.A., Dunbar, R.B., Gonzàlez, J.J., Hayden, T.G., Iwai, M., Jimenez-Espejo, F.J., Katsuki, K., Kong, G.S., McKay, R.M., Nakai, M., Pekar, S.F., Pross, J., Riesselman, C., Sakai, T., Shrivastava, P.K., Tuo, S., Welsh, K., and Yamane, M., 2013. Reorganization of Southern Ocean plankton ecosystem at the onset of Antarctic glaciation. *Science,* 340(6130): 341–344. doi.org/10.1126/science.1223646

Patterson, M.O., McKay, R., Naish, T., Escutia, C., Jimenez-Espejo, F.J., Raymo, M.E., Meyers, S.R., Tauxe, L., Brinkhuis, H., Klaus, A., Fehr, A., Williams, T., Bendle, J.A.P., Bijl, P.K., Bohaty, S.M., Carr, S.A., Dunbar, R.B., Gonzàlez, J.J., Hayden, T.G., Iwai, M., Katsuki, K., Kong, G.S., Nakai, M., Olney, M.P., Passchier, S., Pekar, S.F., Pross, J., Riesselman, C., Röhl, U., Sakai, T., Shrivastava, P.K., Stickley, C.E., Sugisaki, S., Tuo, S., van de Flierdt, T., Welsh, K., and Yamane, M., 2014. Orbital forcing of the East Antarctic ice sheet during the Pliocene and Early Pleistocene. *Nature Geoscience, 7*(11): 841–847. doi.org/10.1038/ngeo2273

Pross, J., Contreras, L., Bijl, P.K., Greenwood, D.R., Bohaty, S.M., Schouten, S., Bendle, J.A., Rohl, U., Tauxe, L., Raine, J.I., Huck, C.E., van de Flierdt, T., Jamieson, S.S.R., Stickley, C.E., van de Schootbrugge, B., Escutia, C., Brinkhuis, H., Welsh, K., McKay, R., and the IODP Expedition 318 Scientists, 2012. Persistent near-tropical warmth on the Antarctic continent during the early Eocene epoch. *Nature*, 488: 73–77. doi.org/10.1038/nature11300

Stocchi, P., Escutia, C., Houben, A. J. P., Vermeersen, B. L. A., Bijl, P. K., Brinkhuis, H., Deconto, R. M., Galeotti, S., Passchier, S., Pollard, D., Brinkhuis, H., Escutia, C., Klaus, A., Fehr, A., Williams, T., Bendle, J. a. P., Bijl, P. K., Bohaty, S. M., Carr, S. A., Dunbar, R. B., Flores, J. A., Gonzàlez, J. J., Hayden, T. G., Iwai, M., Jimenez-Espejo, F. J., Katsuki, K., Kong, G. S., Mckay, R. M., Nakai, M., Olney, M. P., Passchier, S., Pekar, S. F., Pross, J., Riesselman, C., Röhl, U., Sakai, T., Shrivastava, P. K., Stickley, C. E., Sugisaki, S., Tauxe, L., Tuo, S., Van De Flierdt, T., Welsh, K., and Yamane, M., 2013. Relative sea-level rise around East Antarctica during Oligocene glaciation. *Nature Geoscience*, 6: 380–384. doi.org/10.1038/ngeo1783

Great Barrier Reef Environmental Changes: IODP Expedition 325

Jody M. Webster, Geocoastal Research Group,
School of Geosciences, University of Sydney

Introduction

IODP Expedition 325, 'Great Barrier Reef Environmental Changes', which investigated the fossil shelf-edge reefs of the Great Barrier Reef (GBR), was the fourth IODP expedition to use a mission-specific platform, and was conducted by the ECORD Science Operator (ESO) for the European Consortium for Ocean Research Drilling (ECORD). The scientific objectives were to establish the course of sea-level change, define sea-surface temperature (SST) variations and analyse the impact of these environmental changes on reef growth and geometry over the period of 20–10 ka (thousands of years ago). Expedition 325 complemented and extended the findings of the 2005 Expedition 310, 'Tahiti Sea Level', which recovered postglacial coral reef cores from the flanks of Tahiti from 41.6–117.5 m below sea level (mbsl) and spanned ~16 to ~8 ka.

Operational results

During Expedition 325, 34 holes were cored from 17 sites (M0030–M0058) at three locations (Hydrographer's Passage, Noggin Pass and Ribbon Reef) in water depths between 42 and 157 mbsl from the research vessel *Greatship Maya* (Webster et al., 2011) (Figure 7.9). Locations are distributed along the margin to assess the impact of regional variations in oceanographic conditions, SST, sediment input, shelf-edge morphology (width and slope) and glacialhydro-isostatic behaviour on reef response. The drilling strategy focused on recovering fossil coral reef deposits from the Last Glacial Maximum (LGM) to 10 ka. This was achieved by drilling transects of holes through the most prominent fossil barrier reef structures between 40–50 mbsl and the series of well-developed reef terraces between 80–130 mbsl (Figure 7.9, bottom panel).

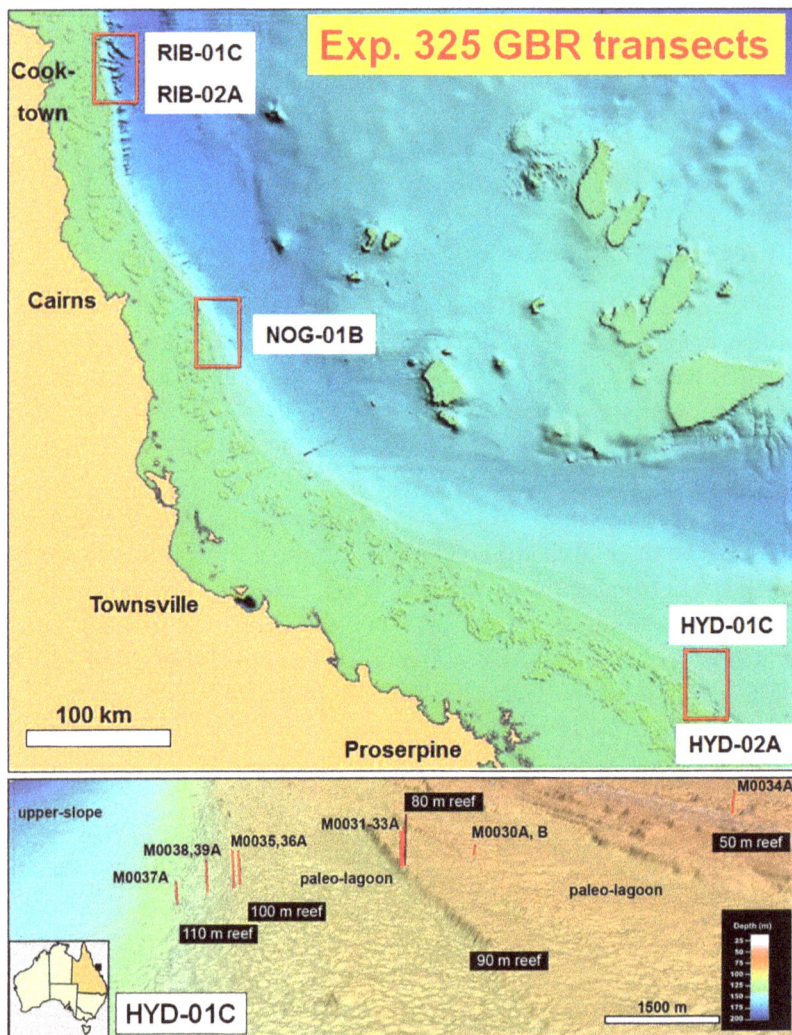

Figure 7.9. Location maps showing transects, and key sites on one transect

The upper panel is a map showing the location of the IODP Expedition 325 transects (after Camoin and Webster, 2014). The bottom panel shows a close-up 3D bathmetric view of Expedition 325 transect HYD-01C and sites M0030A–38A (after Yokoyama et al., 2011 and Abbey et al., 2011).

Source: Yokoyama et al. (2011)

Two transects were drilled at Hydrographer's Passage: a northern transect (HYD-01C), which includes Sites M0030–39, and a southern transect (HYD-02A) encompassing Sites M0041–48. In the Ribbon Reef region, only four sites (M0049–51) from the southern RIB-02A transect were drilled. However, at Noggin Pass another transect (NOG-01B) consisting of M0052–58 – including a hole (M0058A) on the upper continental slope (Harper et al., 2015; Herrero-Bervera and Jovane, 2013) – was completed. Challenging drilling conditions (serious technical and weather difficulties, unconsolidated sediments, cavities, etc.) meant that average per cent recoveries (27.2 per cent: Yokoyama et al., 2011) were lower than Expedition 310 (57.5 per cent: Camoin et al., 2012; Yokoyama et al., 2011), according to standard IODP calculations. However, several strategies were employed over time to maximise core recovery and quality including shorter cores runs (1 m), drilling closely spaced (within 10 m) replicate holes to generate composite cores and, most importantly, the successful implementation of the HQ drilling string at the last Sites M0054–57 saw average recoveries increase to >40–50 per cent. Like Tahiti, efforts were made at several sites (M0042A, 55A, 56A, 57A) to drill deeper into the older Pleistocene deposits to understand the nature of the pre-LGM substrate and to provide new information about reef development, diagenetic environments and sea level prior to this period (i.e. Gischler et al., 2013).

Initial scientific results

At the time of writing this summary, the majority of Expedition 325 post-cruise analyses have been completed and their publication is still in progress. However, a synthesis (Camoin and Webster, 2014) of the initial results from the proceedings, published site survey data and the first papers already confirms that Expedition 325 represents an important new record of sea level, environmental changes and reef response over the last 30 ka.

Shelf-edge reef chronostratigraphy

Sixty-eight fossil coral samples were dated by U/Th and 14C-AMS from representative cores from the top, middle and bottom of the Expedition 325 sites. This was undertaken to provide a basic chronostratigraphy for the drowned GBR shelf edge reefs so as to better guide the sampling party and post-cruise analyses (see Webster et al., 2011 for details). These preliminary data suggest that the reefs are composed of two basic chronostratigraphic sequences (Figure 7.10): basal >MIS3 (~30 ka) deposits and the overlying MIS2 to last deglacial coral reef deposits (Camoin and Webster, 2014). Below the inner barrier at HYD-01C (Hole M0034A) and NOG-01B (Hole M0057A) and inner terrace at NOG-01B (Hole M0057A), the

>MIS3 deposits are clearly reefal, and diagenetic evidence (e.g. dissolution, brownish staining) suggests they have been subaerially exposed (Gischler et al., 2013) prior to reflooding, reef initiation and growth during the last deglacial. In contrast, the >MIS3 deposits below the deeper terraces (90–110 mbsl) (e.g. Holes M0031–39A, 43A, 55A, 53A, 54A, 54B) are composed of dark grainstones and packstones characterised by shells, coral, coralline algae, Halimeda and abundant larger benthic foraminifera representing lower shelf/slope settings (Webster et al., 2011). The contact between the two sequences represents a major unconformity surface and has also been recognised in the downhole and sample petrophysical data (Webster et al., 2011; Yokoyama et al., 2011), and mapped regionally as well-defined seismic reflectors (Hinestrosa et al., 2014).

Composition of the MIS2 to deglacial reef sequence

The MIS2 to deglacial coral reef deposits (~30–10 ka) are composed mainly of coral reef frameworks and detrital sedimentary facies (Figure 7.10). Three boundstone facies are defined based on their varying proportions of corals, coralline algae and microbial sediments. In these framework facies, coral growth forms include massive, robust branching, branching, tabular, encrusting and foliaceous, and they are commonly encrusted by thick centimetre-scale layers of coralline algae, encrusting foraminifera and associated vermetid gastropods (Webster et al., 2011). While they are not as ubiquitous as in Expedition 310 (Seard et al., 2011), some intervals within the deeper terraces, particularly at NOG-01B (Holes M0053A, 54A, 54B), are dominated by abundant microbialite crusts exhibiting complex laminated and thrombolitic morphologies. The coralgal to microbialite dominated boundstones are also associated with abundant consolidated and unconsolidated sediments that are composed of mollusks, benthic foraminifera, red algae and bryozoans that occur locally as internal sediments or as thick (1–19 m) intervals underlying the boundstone facies (e.g. Holes M0031–33A) (Figure 7.10). The most common facies succession includes unconsolidated sediments at the base of the cores overlying the >MIS3 lower shelf/slope sequence. These sediments then grade upward into 20–30 m thick framework-dominated intervals composed of coralgal-microbialite to coralgal boundstone deposits forming the MIS2 to deglacial reef sequence (Figure 7.10).

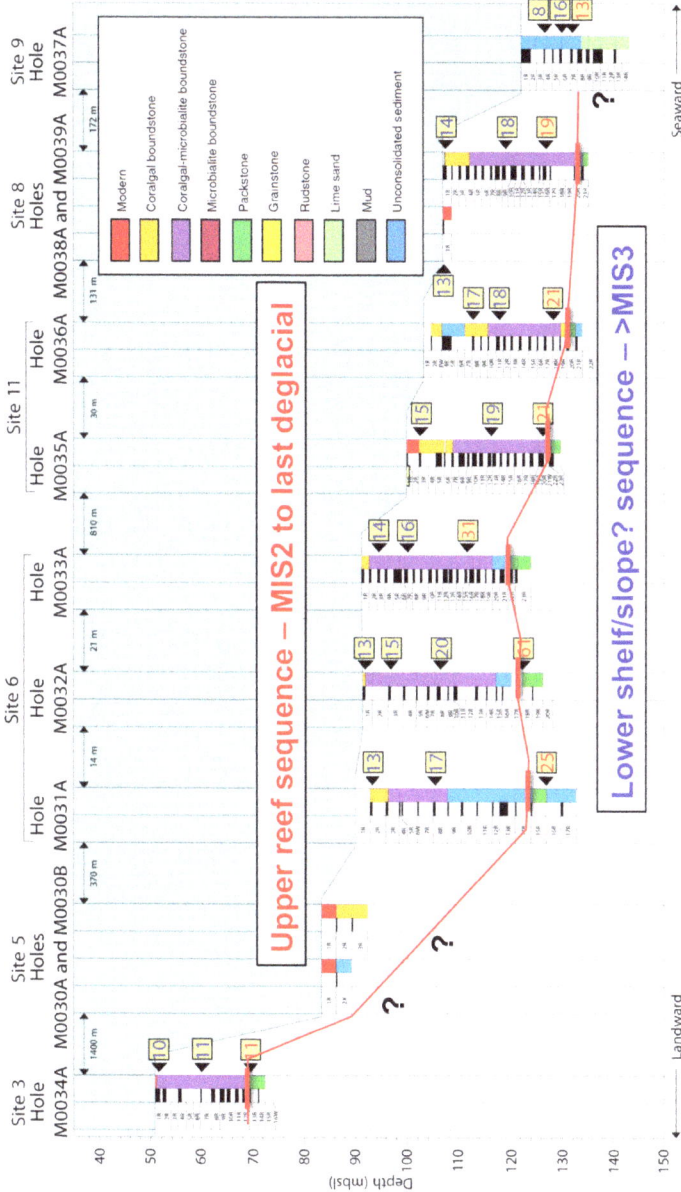

Figure 7.10. IODP Expedition 325 transect HYD-01C showing basic facies patterns and age structure defining the two main sedimentary sequences (after Webster et al., 2011)

The numbers in boxes to the right of the stratigraphic columns represent preliminary core catcher U/Th (red) and C14-AMS ages.

Source: Webster et al. (2011); Camoin and Webster (2014)

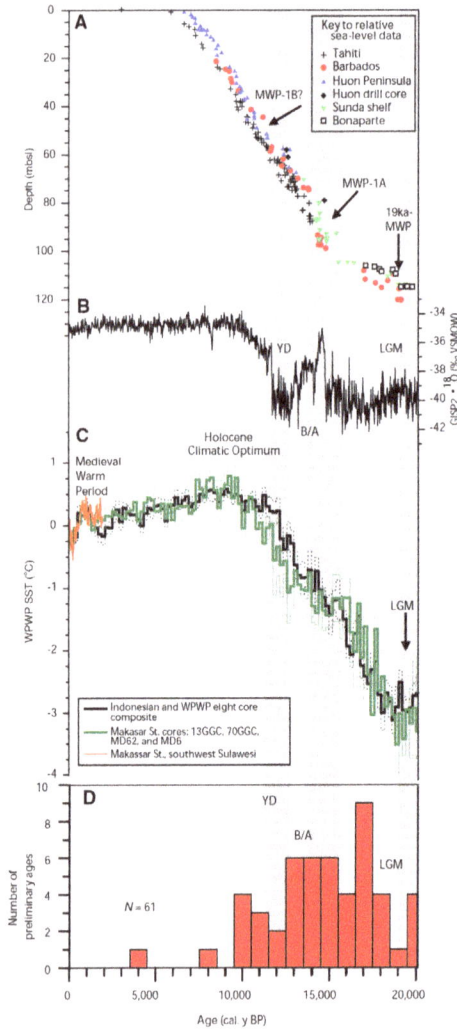

Figure 7.11. Comparison of preliminary dating results from Expedition 325 with previously published sea level and palaeoclimate data

A. Previously published relative sea-level data from Tahiti, Barbados, Huon Peninsula, Bonaparte Gulf and the Sunday shelf. B. GISP2 δ18O proxy for temperature over Greenland. C. Sea-surface temperature variation in the Western Pacific Warm Pool (WPWP). D. Histogram showing the initial U/Th and C14-AMS dating results from the core catcher samples (Webster et al., 2011). The age distribution indicates that the recovered fossil coral reef cores cover key intervals of interest for sea-level changes and environmental reconstruction, including the Last Glacial Maximum (LGM), Bölling-Alleröd warming (B/A), and Younger Dryas (YD). Note that more than 850 new dates have since been generated by the Expedition 325 dating teams.

Source: After Yokoyama et al., 2011 and see this reference for all original data sources

In the Expedition 325 cores, coral assemblages are dominated by massive Isopora, robust branching Acropora and branching Seriatopora, but massive Porites and Faviidae, encrusting Porites, and Montipora with foliaceous Agariciids are locally abundant. Hydrolithon onkodes is the most abundant coralline algae, together with Lithophyllum prototypum and Neogoniolithon fosliei (Webster et al., 2011; Yokoyama et al., 2011). Comparison with their modern environments in the GBR suggests that these assemblages are characteristic of shallow reef crest to deeper reef slopes, and are consistent with the reconstructed Expedition 310 environments (Abbey et al., 2011; Camoin et al., 2012) and other studies of Indo-Pacific reef systems (Montaggioni, 2005).

Preliminary observations confirm that deepening upward coralgal successions are common in the top 3–5 m of the cores. Combined with other sedimentological characteristics (i.e. intense bioerosion, manganese and iron staining) (Webster et al., 2011), this represents a classic reef drowning signature also observed at the top of the Expedition 310 deglacial reef (Abbey et al., 2011; Camoin et al., 2012), and the adjacent GBR dredge samples. For example, Abbey et al. (2013) conducted sedimentologic, palaeoenvironmental and chronologic studies of dredged coral, algae and bryozoan specimens from the tops of the GBR shelf edge reefs. Two distinct generations of fossil mesophotic coral community development are observed between 13–10 ka and 8 ka that have been influenced by widespread, massive flux of siliciclastic sediments associated with the flooding of the GBR shelf during deglacial sea-level rise.

Potential for reconstructing reef growth, sea level and palaeoclimate change

IODP Expedition 325 recovered fossil coral reef deposits from 46 to 145 mbsl, with a preliminary age range of between 9 ka to older than 30 ka. Figure 7.11D shows the distribution of the core catcher ages and their relationship to previously published sea level (Figure 7.11A) and palaeoclimate data (Figures 7.11B, 7.11C) since the LGM (see Yokoyama et al., 2011). This figure highlights the excellent chronologic coverage of the key palaeoenvironmental intervals (LGM, Bölling-Alleröd, Younger Dryas, and Medieval Warm Period (MWP) events) provided by Expedition 325 cores, particularly in the context of the +900 new U/Th and C14-AMS measurements on corals and algae that have now been obtained. Combined with firm palaeowater-depth estimates provided by facies and coralgal analysis and glacial isostatic modelling, these ages will allow the reconstruction of a robust, new sea-level curve from 30 to 10 ka. Numerous massive coral colonies suitable for palaeoclimate studies were also recovered that will help define SST and sea surface salinity variations during this period in the southwest Pacific. For example, based on new stable isotope and Sr/Ca data from the Expedition 325 corals, Felis et al. (2014) reported that the SSTs in the GBR were significantly cooler than previously assumed, and that a larger than expected north–south temperature gradient existed 20–13 ka. Finally, once the precise stratigraphic, chronologic and sea-level framework has been established, 3D numerical reef modelling will allow the investigation of the response of the GBR to major environmental perturbations over the last 30 ka.

References

Abbey, E., Webster, J.M., and Beaman, R.J., 2011. Geomorphology of submerged reefs on the shelf edge of the Great Barrier Reef: The influence of oscillating Pleistocene s. *Marine Geology*, 288: 61–78. doi.org/10.1016/j.margeo.2011.08.006

Abbey, E., Webster, J.M., Braga, J.C., Jacobsen, G.E., Thorogood, G., Thomas, A.L., Camoin, G., Reimer, P.J., and Potts, D.C., 2013. Deglacial mesophotic reef demise on the Great Barrier Reef. *Palaeogeography, Palaeoclimatology, Palaeoecology*, 392: 473–494. doi.org/10.1016/j.palaeo.2013.09.032

Camoin, G.F., Seard, C., Deschamps, P., Webster, J.M., Abbey, E., Braga, J.C., Iryu, Y., Durand, N., Bard, E., Hamelin, B., Yokoyama, Y., Thomas, A.L., Henderson, G.M., and Dussouillez, P., 2012. Reef response to sea-level and environmental changes during the last deglaciation: Integrated Ocean Drilling Program Expedition 310, Tahiti Sea Level. *Geology*, 40: 643–646. doi.org/10.1130/G32057.1

Camoin, G., and Webster, J.M., 2014. Coral reefs and sea level change, in: Stein, R., Blackman, D., Inagaki, F., and Christian-Larsen, H. (eds), *Earth and Life Processes Discovered from Subseafloor Environment: A Decade of Science Achieved by the Integrated Ocean Drilling Program (IODP)*, Volume 7. Elsevier: Amsterdam/New York, p. 822. doi.org/10.1016/b978-0-444-62617-2.00015-3

Felis, T., McGregor, H.V., Linsley, B.K., Tudhope, A.W., Gagan, M.K., Suzuki, A., Inoue, M., Thomas, A.L., Esat, T.M., Thompson, W.G., Tiwari, M., Potts, D.C., Mudelsee, M., Yokoyama, Y., and Webster, J.M., 2014. Intensification of the meridional temperature gradient in the Great Barrier Reef following the Last Glacial Maximum. *Nature Communications*, 5. doi.org/10.1038/ncomms5102

Gischler, E., Thomas, A.L., Droxler, A.W., Webster, J.M., Yokoyama, Y., Schöne, B.R., and Porta, G.D., 2013. Microfacies and diagenesis of older Pleistocene (pre-last glacial maximum) reef deposits, Great Barrier Reef, Australia (IODP Expedition 325): A quantitative approach. *Sedimentology*, 60: 1432–1466. doi.org/10.1111/sed.12036

Harper, B.B., Puga-Bernabéu, Á., Droxler, A.W., Webster, J.M., Gischler, E., Tiwari, M., Lado-Insua, T., Thomas, A.L., Morgan, S., Jovane, L., and Röhl, U., 2015. Mixed Carbonate–Siliciclastic Sedimentation Along the Great Barrier Reef Upper Slope: A Challenge to the Reciprocal Sedimentation Model. *Journal of Sedimentary Research*, 85: 1019–1036. doi.org/10.2110/jsr.2015.58.1

Herrero-Bervera, E., and Jovane, L., 2013. On the palaeomagnetic and rock magnetic constraints regarding the age of IODP 325 Hole M0058A, in Jovane, L., Herrero-Bervera, E., Hinnov, L. A. & Housen, B. A. (eds), Magnetic Methods and the Timing of Geological Processes. *Geological Society, London, Special Publications*, London, pp. 279–291. doi.org/10.1144/sp373.19

Hinestrosa, G., Webster, J.M., Beaman, R.J., and Anderson, L.M., 2014. Seismic stratigraphy and development of the shelf-edge reefs of the Great Barrier Reef, Australia. *Marine Geology*, 353: 1–20. doi.org/10.1016/j.margeo.2014.03.016

Montaggioni, L.F., 2005. History of Indo-Pacific coral reef systems since the last glaciation: Development patterns and controlling factors. *Earth-Science Reviews*, 71: 1. doi.org/10.1016/j.earscirev.2005.01.002

Seard, C., Camoin, G., Yokoyama, Y., Matsuzaki, H., Durand, N., Bard, E., Sepulcre, S., and Deschamps, P., 2011. Microbialite development patterns in the last deglacial reefs from Tahiti (French Polynesia; IODP Expedition #310): Implications on reef framework architecture. *Marine Geology*, 279: 63–86. doi.org/10.1016/j.margeo.2010.10.013

Webster, J.M., Yokoyama, Y., Cotterill, C., Anderson, L., Green, S., Bourillot, R., Braga, J.C., Drowler, A., Esat, T., Felis, T., Fujita, K., Gagan, M., and the Expedition 325 Scientists, 2011. *Great Barrier Reef environmental changes. Proceedings of the IODP*, 325: Tokyo (Integrated Ocean Drilling Program Management International, Inc.). doi.org/10.2204/iodp.proc.325.2011

Yokohama, Y., Webster, J.M., Cotterill, C., Braga, C., Jovane, L., Mills, H., Morgan, S., Suzuki, A., Anderson, L., Green, S., Bourillot, R., Drowler, A., Esat, T., Felis, T., Fujita, K., Gagan, M., and the IODP Expedition 325 Scientists, 2011. IODP Expedition 325: The Great Barrier Reef reveals past sea-level, climate, and environmental changes since the last Ice Age. *Scientific Drilling*, 12: 32–45. doi.org/10.2204/iodp.sd.12.04.2011

A (wandering?) tail of two plumes, Louisville and Hawaii: IODP Expedition 330

Benjamin Cohen, then University of Queensland,
now University of Glasgow

Even a cursory look at a map of the planet reveals that the ocean floor is littered with underwater mountains – with over 100,000 such 'seamounts' above 1 km high at last estimate (Wessel et al., 2010). The most striking of these underwater mountains form long semi-linear chains extending thousands of kilometres across the ocean basins (Figure 7.12), with the volcanoes becoming progressively older in the direction of plate motion (McDougall, 1964). In the early days of plate tectonic theory, the idea developed that these volcanic chains represent the surface expressions of mantle plumes – tails of buoyant hot material rising from deep within the planet, causing volcanism where the hot rock reaches the surface (Wilson, 1963). If such plumes remain fixed within the mantle, then the volcanoes formed above them should act like a tape recorder, tracking the speed and direction of plate motions back through geologic time (Morgan, 1971; Sharp and Clague, 2006). The archetypical example of hotspot volcanoes recording a change in plate motion is the Hawaiian-Emperor chain, which has a prominent bend at approximately 50 Ma (Figure 7.12).

In the years following the development of plume theory, however, it was argued that plumes may not be fixed, but instead could be bent by changes in mantle convection currents, also known as the 'mantle wind' (Tarduno et al., 2009). Knowledge of the origin – and fixity – of hotspot volcanoes and underlying purported plumes is crucial, not only to understand the formation of thousands of seamounts scattered across the ocean basins, but also because the time–space distribution of hotspot volcanoes provides an important reference frame in reconstructing past motions of Earth's plates back through geological history.

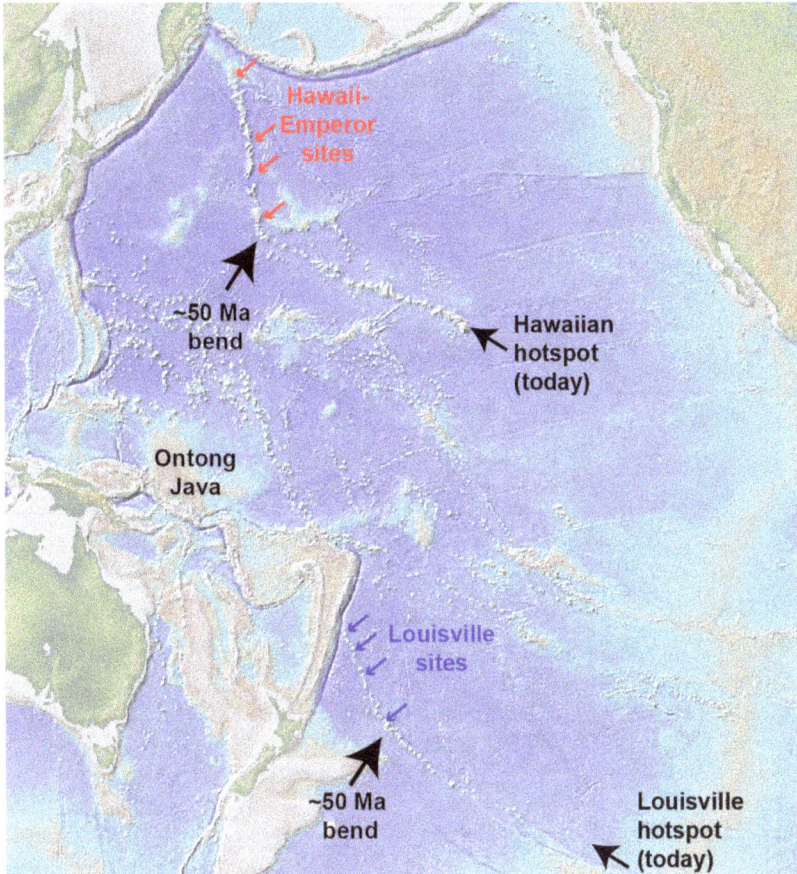

Figure 7.12. Bathymetric map of the Pacific Ocean, showing abundant undersea mountains (seamounts)

The Hawaiian-Emperor and Louisville volcanic chain both have bends at ~50 Ma, although the Hawaiian-Emperor bend is more pronounced.

Source: Image modified from GeoMapApp

Deep-sea drilling has been instrumental to test the fixity of mantle plumes. The first investigation focused on the Hawaiian-Emperor chain (Tarduno et al., 2003), with four seamounts drilled during DSDP Leg 55 and ODP Leg 197 (Figure 7.12). A highlight of these expeditions was the palaeomagnetic measurements, which revealed that the Hawaiian-Emperor plume had moved by ~15 degrees of latitude prior to 50 Ma (Figure 7.13). Thus it appears that the Hawaiian plume has wandered in the past, although recent analysis (Whittaker et al., 2007) also calls for change in plate motion at 50 Ma.

111

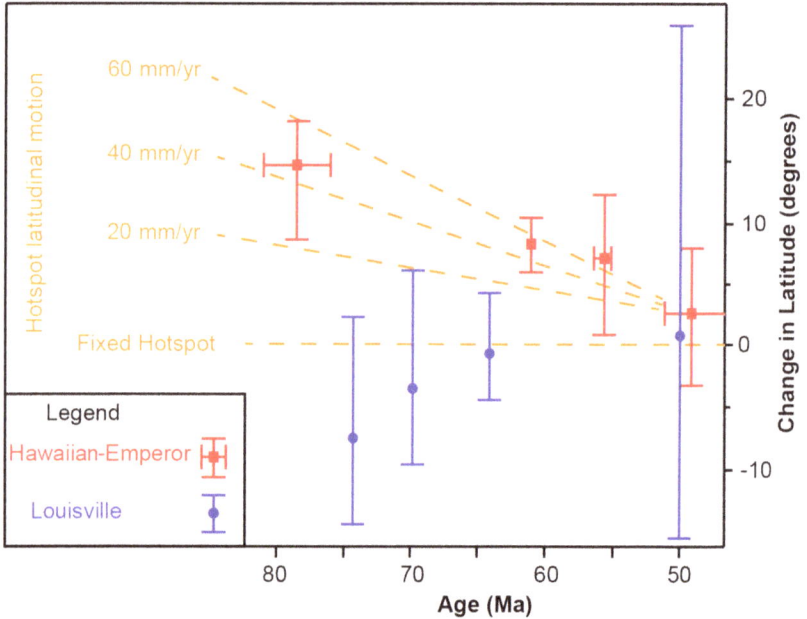

Figure 7.13. Palaeomagnetic results from deep-sea drilling of two hotspot chains in the Pacific

Results from Hawaiian-Emperor chain indicate latitudinal motion of up to ~15 degrees, but there is minimal latitudinal hotspot motion for the Louisville chain.

Source: Modified after Koppers et al. (2012) and Tarduno et al. (2003)

The Hawaiian-Emperor plume is but one example. Had other plumes also been influenced by mantle convection? To find out, on December 2010, the IODP ship *JOIDES Resolution* set out to drill the Louisville seamounts, in the southwest Pacific Ocean (Figure 7.12). Drilling of the older (northern) end of the Louisville chain enabled a direct temporal comparison between two plumes (i.e. between ~80 and 50 Ma; Figure 7.13), but on opposite ends of the Pacific Plate. The Louisville expedition (IODP 330) was highly successful, recovering over 800 m of remarkably fresh volcanic and sedimentary rocks, at an average recovery of 72.4 per cent – an ocean drilling record for hard-rock recovery. Palaeomagnetic measurements were also successful but, in contrast to Hawaii, results from Louisville are within analytical error of the modern-day hotspot location, indicating minimal latitudinal hotspot motion (Figure 7.13). Thus it appears that Louisville and Hawaii are moving independently, and are not influenced by the same convection systems in the mantle. The Louisville data also rule out true polar wander as the cause of the Hawaiian palaeomagnetic results.

In addition to this question of hotspot fixity, the scientists aboard were also carrying out a multidisciplinary range of research, including measuring the longevity and eruptive rates at the volcanoes, investigating the links between the Louisville chain and the Ontong Java Plateau (the world's largest igneous province – covering an area comparable to the size of Greenland), the remarkable geochemical homogeneity of the Louisville lavas, the palaeoclimate record of sediments and fossils on top of the volcanoes, the amount of CO_2 sequestered by seawater alteration of the volcanic rocks, and the presence of microbiological activity many hundreds of metres below the sea floor (mbsf) (e.g. Figures 7.14A–D). As such, the rocks obtained from this expedition will provide a valuable scientific resource and enduring legacy, with various investigations set to continue years into the future.

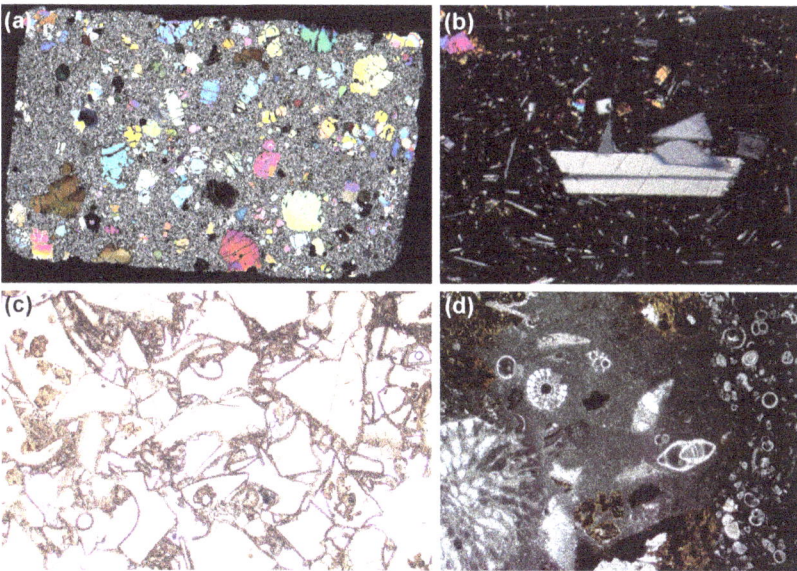

Figure 7.14. A selection of other scientific highlights from IODP 330, including a variety of exceptionally fresh volcanic rocks

(A) fresh olivine and augite phyric lava (image 25 mm long) suitable for a variety of geochemical analyses; (B) fresh plagioclase (in this case 1 mm long with a distinctive JR ship-shaped outline), which is suitable for geochronology; and (C) fresh glass, which is ideal for chemical analyses, and for trapping volatile volcanic gases (image ~5 mm wide). The dark edges on the glass are numerous tubular features formed by microorganisms. (D) The expedition also recovered fossil-bearing sediments that provide a biological and climatic record for the southwest Pacific (image 5 mm long).

Source: Ben Cohen

The future of seamount exploration is very promising. Exciting targets remain – especially in the Atlantic and Indian Oceans – providing ideal sites for the International Ocean Discovery Program to continue the path set by the *Glomar Challenger* in the early years of scientific drilling.

References

Koppers, A.A.P., Yamazaki, T., Geldmacher, J., Gee, J.S., Pressling, N., Hoshi, H., Anderson, L., Beier, C., Buchs, D.M., Chen, L.-H., Cohen, B.E., Deschamps, F., Dorais, M.J., Ebuna, D., Ehmann, S., Fitton, J.G., Fulton, P.M., Ganbat, E., Hamelin, C., Hanyu, T., Kalnins, L., Kell, J., Machida, S., Mahoney, J.J., Moriya, K., Nichols, A.R.L., Rausch, S., Sano, S.-i., Sylvan, J.B., and Williams, R., 2012. Limited latitudinal mantle plume motion for the Louisville hotspot. *Nature Geoscience,* 5: 911–917. doi.org/10.1038/ngeo1677

Koppers, A.A.P., Yamazaki, T., Geldmacher, J., Anderson, L., Beier, C., Buchs, D.M., Chen, L.-H., Cohen, B.E., and the IODP Expedition 330 Scientists, 2012. *Proceedings of the IODP,* 330: Tokyo (Integrated Ocean Drilling Program Management International, Inc.).

McDougall, I., 1964. Potassium-Argon Ages from Lavas of the Hawaiian Islands. *Geological Society of America Bulletin,* 75: 107–128. doi.org/10.1130/0016-7606(1964)75[107:PAFLOT]2.0.CO;2

Morgan, W.J., 1971. Convection Plumes in the Lower Mantle. *Nature,* 230: 42–43. doi.org/10.1038/230042a0

Morgan, W.J., 1972. Deep Mantle Convection Plumes and Plate Motions. *The American Association of Petroleum Geologists Bulletin,* 56: 203–213.

Sharp, W.D., and Clague, D.A., 2006. 50-Ma Initiation of Hawaiian-Emperor Bend Records Major Change in Pacific Plate Motion. *Science,* 313: 1281–1284. doi.org/10.1126/science.1128489

Tarduno, J., Bunge, H.-P., Sleep, N.H., and Hansen, U., 2009. The Bent Hawaiian-Emperor Hotspot Track: Inheriting the Mantle Wind. *Science,* 324: 50–53. doi.org/10.1126/science.1161256

Tarduno, J.A., Duncan, R.A., Scholl, D.W., Cottrell, R.D., Steinberger, B., Thordarson, T., Kerr, B.C., Neal, C.R., Frey, F.A., Torii, M., and Carvallo, C., 2003. The Emperor Seamounts: Southward Motion of the Hawaiian Hotspot Plume in Earth's Mantle. *Science,* 301: 1064–1069. doi.org/10.1126/science.1086442

Wessel, P., Sandwell, D.T., and Kim, S.-S., 2010. The Global Seamount Census. *Oceanography* 23: 24–33. doi.org/10.5670/oceanog.2010.60

Whittaker, J.M., Müller, R.D., Leitchenkov, G., Stagg, H., Sdrolias, M., Gaina, C., and Goncharov, A., 2007. Major Australian-Antarctic Plate Reorganization at Hawaiian-Emperor Bend Time. *Science* 318: 83–86. doi. org/10.1126/science.1143769

Wilson, J.T., 1963. A possible origin of the Hawaiian Islands. *Canadian Journal of Physics* 41: 863–870. doi.org/10.1139/p63-094

The Japan Trench Rapid Drilling Project (JFAST) yields new insights into the mechanics and structure of subduction thrust faults: IODP Expeditions 343 and 343T

Virginia G. Toy, University of Otago, Dunedin

(Expedition 343 and 343T Scientists, c/- CDEX, JAMSTEC, 3173-25 Showa-machi, kanazawa-ku, Yokohama Kanagawa 236-0001, Japan)

The 2011 Mw9.0 Tohoku-oki earthquake ruptured the Japan Trench, with a very large coseismic slip occurring on the shallow part of the décollement. A significant consequence of this large slip was the generation of a devastating tsunami that impacted coastal Sendai. To better understand the controls on rupture propagation and slip, the plate boundary décollement and over-riding and subducting plate materials (Figure 7.15) near the trench were investigated by downhole logging and coring, and a temperature observatory was installed during IODP Expeditions 343 and 343T (the Japan Trench Fast Drilling Project (JFAST)) from April to July 2011. The onboard science party of 34 included experts from 11 countries.

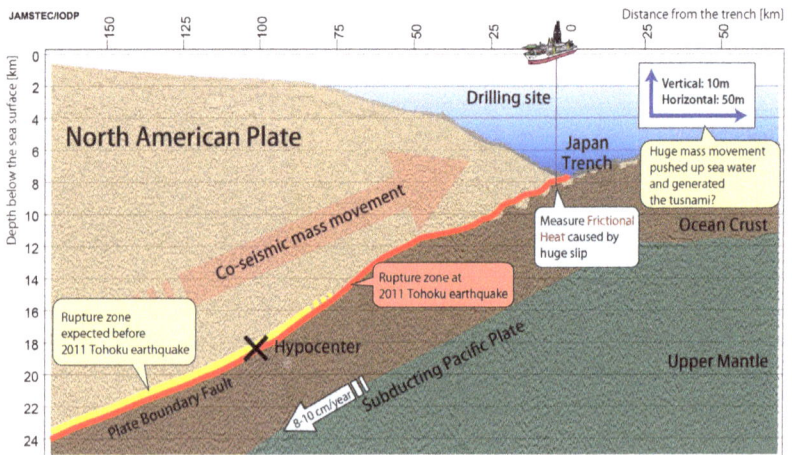

Figure 7.15. Schematic diagram illustrating the structure of the accretionary prism at the Japan Trench sampled and monitored during the JFAST expeditions

Source: Professor Jim Mori, Kyoto University

JFAST set a number of records in scientific ocean drilling. Major project achievements include drilling the world's deepest research borehole below sea level (6,889.5 m water depth + 850.5 m borehole = 7,740 m below sea level (mbsl); the previous record was 7,049.5 mbsl during DSDP Leg 60, Hole 461A in the Marianas Trench in 1978, where in a water depth of 7,034 m the project achieved 15.5 m penetration below the sea floor). The JFAST expeditions recovered the deepest oceanic research core (7,734 mbsl), sampled the Japan Trench plate boundary décollement (subduction thrust fault) between Miocene and Cretaceous sediments for the first time, and installed and recovered the first temperature observatory across an active plate boundary fault soon enough after it experienced earthquake-generating slip to measure the thermal perturbation resulting from frictional heating during the slip.

Analysis of samples and data arising from Expeditions 343 and 343T has already yielded new insights into subduction thrust fault mechanics. Significant research results published so far include the following:

- Borehole breakouts that were imaged downhole (Lin et al., 2013) indicate the hanging wall accretionary prism above the décollement changed to an extensional stress regime after the earthquake. This confirms seismological suggestions that the shallow parts of the fault accomplished near total stress release during the 2011 event. It is more generally found that faults release only a small proportion of the stress resolved at their earthquake foci (e.g. Baltay et al., 2011).

- Part of the plate boundary décollement was sampled in core (Chester et al., 2013; Figure 7.16). It is revealed to be a narrow (<5 m), localised sheared zone comprising scaly phaccoids that have striated lustrous surfaces and asymmetry consistent with the sense of shear on modern plate boundary. Lithologically, the décollement material was originally volcanigenic sediment, but it has been extensively authigenically altered. Mineralogical analyses using X-ray diffraction reveal it is particularly rich in the clay mineral smectite (78 per cent; for comparison Nankai trough plate boundary rocks have 31 per cent smectite).

Figure 7.16. Drs Marianne Cronin and Kohtaro Ujiie and Professor Jim Mori (PI) examine core '17R', which sampled part of the plate boundary décollement between the Japan and Eurasian plates at the Expedition 343 site

Source: Expedition 343 online photo gallery, www.jamstec.go.jp/chikyu/e/exp343/gallery.html

- Frictional experiments have been performed at low to high (1.3 m/s) velocities on décollement materials (Ujiie et al., 2013). These indicate the weak, foliated clay has a very low friction coefficient (μk = 0.19 if fully drained or μk = 0.03 if placed between impermeable forcing blocks so thermal pressurisation occurs during slip). Foliated and unfoliated gouges and injection structures generated in the experiments are distinctive microstructures that can be compared to natural faults to help us predict their behaviour.

- The temperature observatory data (Fulton et al., 2013; Figure 7.17) contain a thermal anomaly of 0.31°C at the décollement. This indicates that a maximum temperature of 1250°C was attained during slip and 27 megajoules/square metre of energy was dissipated. Thus coseismic frictional strength was very low (μk = 0.08), consistent with experimental data if some thermal pressurisation did occur. Careful modelling of how thermal perturbation can occur in rock suggests the anomaly at the depth of the décollement is not a result of flow of hot fluid up the fault, but such fluid flow does explain thermal perturbations around more steeply dipping faults shallower in the prism.

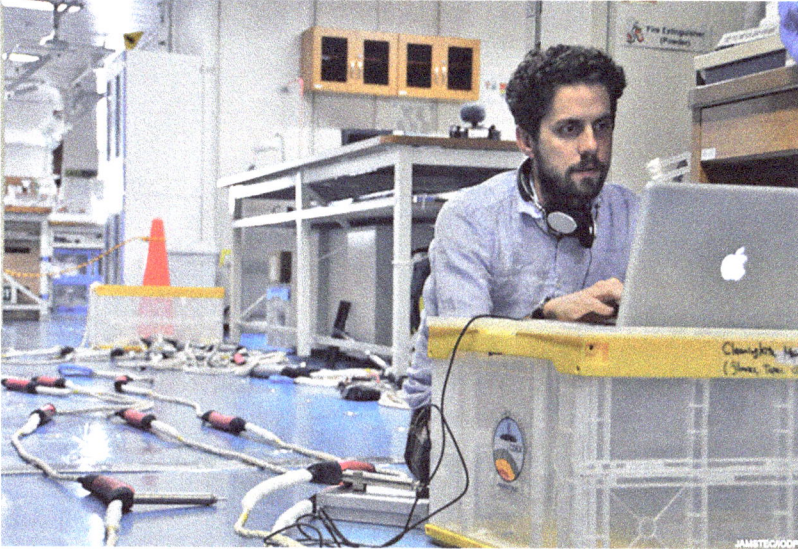

Figure 7.17. A slightly harrowed-looking Dr Patrick Fulton of University of California, Santa Cruz, redesigns the temperature observatory one last time (honest!) as the exact depth of the predicted fault intercept changes

The temperature sensors, attached to a rope at carefully designed spacings, are arrayed on the floor behind Patrick.

Source: Expedition 343 online photo gallery, www.jamstec.go.jp/chikyu/e/exp343/gallery. html

To date, New Zealand–based researchers and their collaborators have contributed to understanding of the plate boundary décollement by:

1. Making measurements of particle size and shape developed in two contrasting lithologies by drilling-induced fragmentation (Toy et al., 2013a). These data will be related to drilling parameters in an attempt to quantify energy input for comparison with the results of natural fragmentation processes.

2. Study of development of fabrics throughout the prism, in collaboration with Dr Ake Fagereng then at the University of Cape Town, South Africa, and now at Cardiff University in Wales (Toy et al., 2013b). We are trying to address how the fabrics within the scaly clay of the décollement form: are they just a highly evolved version of the lithologically defined lozenges observed in the hanging wall accretionary prism, or do they specifically form in materials with the particularly clay-rich composition of the décollement? What is the relative timing of alteration of the incoming sediments to clay and development of the fabrics? We are combining thin

section observations of hanging wall and foot wall material with tomographic images obtained via shipboard (core scale), micro- and even synchrotron-scale tomography that reveal the three-dimensional geometry of the structures. Exciting recent thin section observations of the plate boundary décollement materials clearly illustrate that the microscale clay fabric is aligned around many lozenge margins; thus its structure is likely very important in imposing weakness that facilitates shear localisation and large coseismic slip.

In summary, the very smectite-rich shallow décollement materials are demonstrated experimentally, and by the very small temperature anomaly generated during slip, to be extremely frictionally weak (μ~0.05–1.0) due to a combination of unique mineralogy and fabric. They could also have experienced thermal pressurisation resulting in additional coseismic weakening. This allowed an energetic rupture that had initiated deeper on the subduction thrust to accomplish a very large slip in the surface during a near total stress-drop earthquake. The described composition, fabric and mechanical properties will inform interpretation of the potential of other subduction thrusts to accommodate very large slips, provided we can also sample them – a good target for future IODP expeditions.

References

Baltay, A., Ide, S., Pieto, G., and Beroza, G., 2011. Variability in earthquake stress drop and apparent stress. *Geophysical Research Letters,* 38(6): L06303. doi.org/10.1029/2011GL046698

Chester, F., Mori, J., Eguchi, N., Toczko, S., and the Expedition 343 and 343T Scientists, 2012. Integrated Ocean Drilling Program Expedition 343/343T Preliminary Report, Japan Trench Fast Drilling Project (JFAST). *Proceedings of the IODP.*

Chester, F.M., Rowe, C., Ujiie, K., Kirkpatrick, J., Regalla, C., Remitti, F., Moore, J.C., Toy, V.G., Wolfson-Schwher, M., Bose, S., Kameda, J., Mori, J.J., Brodsky, E.E., Eguchi, N., Toczko, S., and the Expedition 343 and 343T Scientists, 2013. Structure and composition of the plate-boundary slip-zone for the 2011 Tohoku-oki earthquake. *Science,* 342: 1208–1212.

Fulton, P.M., Brodsky, E.E., Kano, Y., Mori, J., Chester, F., Ishikawa, T., Harris, R.N., Lin, W., Eguchi, N., Toczko, S., and the Expedition 343/343T and KR13-08 Scientists, 2013. Low coseismic friction on the Tohoku-oki fault determined from temperature measurements. *Science,* 342: 1214–1217.

Lin, W., Conin, M., Moore, J.C., Chester, F.M., Nakamura, Y., Mori, J.J., Anderson, L., Brodsky, E., Eguchi, N., and the Expedition 343 Scientists, 2013. Stress state in the largest displacement area of the 2011 Tohoku-Oki earthquake. *Science*, 339 (6120): 687–690.

Toy, V.G., Scott, H.R., and the IODP-MI and Expedition 343 Science Party, 2013a. Particle shape and size in drilling-induced breccia. Poster and abstract presented at the Exp. 343 'JFAST' 2nd Post-cruise Meeting, Santa Cruz, CA, 3–6 Sept 2013.

Toy, V.G., Scott, H.R., McNaughton, A., Gessner, K., Fagereng, A., and the IODP-MI and Expedition 343 Science Party, 2013b. Evolution of shear fabrics in accretionary prisms: insights from tomographic data and thin section observations. Poster and abstract presented at the Exp. 343 'JFAST' 2nd Post-cruise Meeting, Santa Cruz, CA, 3–6 Sept 2013.

Ujiie, K., Tanaka, H., Saito, H., Tsutsumi, A., Mori, J.J., Kameda, J., Brodsky, E.E., Chester, F.M., Eguchi, N., Toczko, S., and the Expedition 343/343T Scientists, 2013. Low coseismic shear stress on the Tohoku megathrust determined from laboratory experiments. *Science*, 342: 1211–1214.

A study of the Asian Monsoon in the Japan Sea: IODP Expedition 346

Stephen Gallagher, University of Melbourne

Expedition 346 drilled seven sites in the Sea of Japan/East Sea and two closely spaced sites in the East China Sea in August and September 2013 (Figure 7.18). In total, this expedition recovered 6,135.3 m of core – a record amount to be recovered by any single expedition during IODP. The drilling yielded an unparalleled archive of atmosphere–ocean linkages relating to the East Asian monsoonal system. Sediment in the Sea of Japan/East Sea was last investigated by scientific ocean drilling during ODP Legs 127 and 128, nearly 25 years ago. Expedition 346 was the first scientific drilling expedition to focus exclusively on the climate system in this area. With the East Asian Monsoon directly affecting the water supply of one-third of the global population, the outcomes of this expedition will have a direct bearing on society's understanding of this hugely important and complex atmosphere–ocean climate system.

Objectives addressed by drilling the Sea of Japan/East Sea and East China Sea

1. Determining the timing of onset of variability of the East Asian Summer Monsoon and East Asian Winter Monsoon and their relationship with the variability of jet stream circulation.

Dark and light layers of the deep-sea sediment in the Sea of Japan/East Sea represent changes in the intensity of East Asian Winter Monsoon rainfall in South China. Today, this monsoon produces rain that covers thousands of kilometres of southeast Asia, and is a major source of heat to drive planetary atmospheric circulation. Any study that improves the geohistorical knowledge of this system will improve our understanding of global climate and its possible variability with future climate change. It is possible to trace centimetre- to metre-scale (and often millimetre-scale) dark and light layers across hundreds of kilometres, suggesting that the Sea of Japan/East Sea responded as a single system to climatic and/or oceanographic change. These dark and light layers started at ~2.6 million years ago (Ma) and became more frequent and distinct from ~1.2 Ma to the present. This suggests the variability and intensity of this important monsoonal system increased at this time.

The presence of ice-rafted debris and the occurrence of deep-water ventilation are also related to the intensity of the East Asian Winter Monsoon. Colour reflectance (L*) of the sediment reflects bottom-water ventilation. Expedition 346 showed that deposition of ice-rafted debris (possibly related to cold outbreaks from Siberia) started at ~3.2 Ma at Site U1422 and ~2.7 Ma at Sites U1423 and U1424, whereas L* increased significantly from ~2 to ~1.5 Ma at all deeper-water sites. These findings provide glimpses of changes in sedimentation that are related to the evolution of East Asian Winter Monsoon behaviour and the geohistory of related cold outbreaks emanating from Siberia and northern hemisphere ice sheet evolution. Finally, deep-sea sediment recovered from the Sea of Japan/East Sea sites yield wind-blown dust, and can be used to document changes in the Asian westerly jet stream.

The East Asian Summer Monsoon is a major climate system related to the Indian monsoon that initiates near the Philippines and extends as far north as the Yangtze River Basin and Japan. It is a major source of precipitation in the region and is related to the amount of snow cover in the Tibetan Plateau.

Sites U1428 and U1429 in the northern East China Sea yielded a continuous sequence of strata covering the last ~0.4 million years, which provides evidence of strong variability in sea–surface temperature and salinity, reflecting Yangtze River discharge and the history of the East Asian Summer Monsoon. This evidence is presently being compared with terrestrial loess and stalagmite climate records, allowing the link between climatic hydrology on the Asian continent (traced through reconstruction of Yangtze River discharge) and atmospheric processes to be established.

2. Reconstructing changes in surface and deep-water circulation and surface productivity in the Sea of Japan/East Sea over the last 5 million years.

Palaeoceanographic studies of the history of ventilation of enclosed or partially enclosed seas like the Sea of Japan are important as they not only tell us about potential variability with future climate change, they also form an important analogue for studying similar fossilised marine systems that are often associated with the source and generation of hydrocarbons and/or mineral deposits. Sedimentary colour reflectance (L*) reflects deep-water ventilation. Continuous deep-sea sedimentary records up to 4 Ma were cored at Site U1422, ~5 Ma at Sites U1423, U1424, and U1426, and ~12 Ma at Sites U1425 and U1430 (Figure 7.18). The darker layers are not well burrowed, suggesting oxygen-poor conditions, whereas the lighter layers are more burrowed, suggesting more oxygen-rich conditions. Some of the dark layers are brownish and rich in microfossils such as diatoms, nannofossils, radiolarians and foraminifers, suggesting high marine surface biological productivity. Orbital-scale dark–light colour cycles appeared at ~2.6 Ma, and millennial-scale dark–light cycles appeared at ~1.2 Ma. Orbital-scale dark–light colour cycles also appeared from ~12 to ~8 Ma at Sites U1425 and U1430.

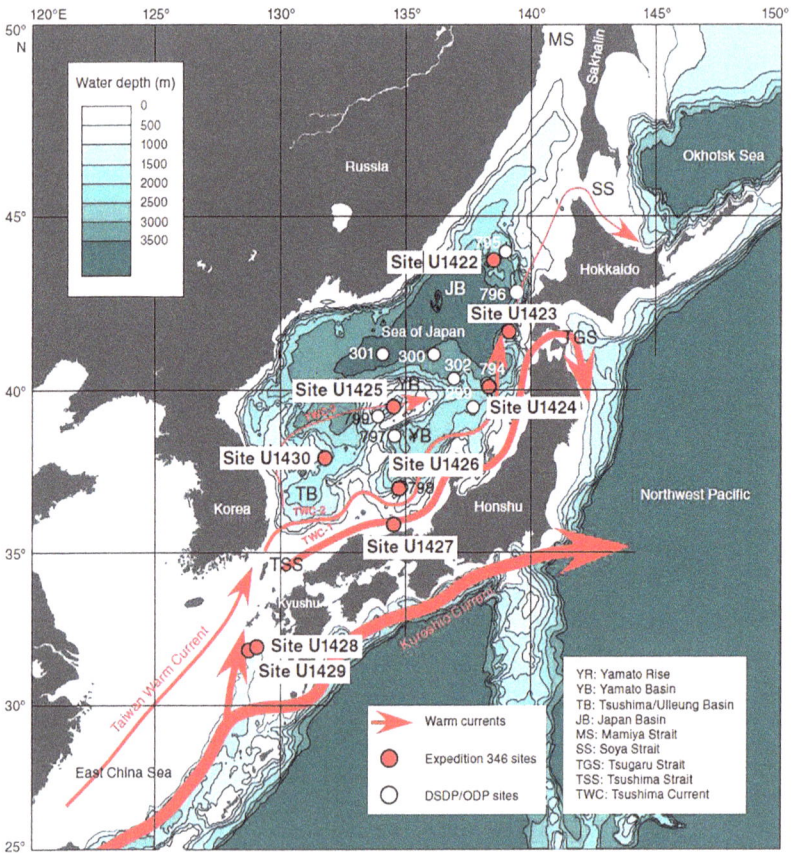

Figure 7.18. Bathymetric map of Expedition 346 sites (red circles) in the Sea of Japan/East Sea and the East China Sea

Sites previously drilled by the Deep Sea Drilling Project (DSDP) and Ocean Drilling Program (ODP) (white circles) are also shown. Also illustrated are surface current systems within and surrounding the Sea of Japan.

Source: Proceedings of the IODP 346, 2015, College Station TX (Integrated Ocean Drilling Program): publications.iodp.org/proceedings/346/346title.htm

8

Education and Outreach

Neville Exon

Education

ECORD Distinguished Lecturers

In 2009, three ECORD (European Consortium for Ocean Research Drilling) Distinguished Lecturers came to Australia and New Zealand under a joint funding arrangement, and gave well-attended talks at various venues. We would like to express our gratitude to ECORD for their support.

Professor Peter Clift of the Department of Geology & Petroleum Geology, University of Aberdeen, gave talks in March at the universities of Sydney, Melbourne and Queensland. The talk was entitled: 'Tibet, the Himalaya and the Development of the Asian Monsoon: A chicken and egg problem for the IODP'.

Professor R. John Parkes, School of Earth and Ocean Science, Cardiff University, gave talks in August at the Australian Microbiological Society Conference and CSIRO in Perth, at the ANU and the universities of Adelaide, Melbourne, Queensland, Waikato and at Victoria University in Wellington. The talk was entitled: 'The prokaryotes and their activities and habitats in sub-seafloor sediments'.

Professor Achim Kopf of the MARUM Research Centre, University of Bremen, gave talks in November to the University of Melbourne and The Australian National University (ANU). The talk was entitled: 'Subduction mega-earthquakes and other geohazards: IODP NanTroSEIZE as a type example for complex scientific drilling'.

IODP Distinguished Lecturer

Professor Ted Moore of the Department of Geological Sciences, University of Michigan, gave two talks at ANU in 2009, and his first visit was co-funded by IODP. His first very well attended public talk in May was entitled: 'Messages from the Past: The warm Earth we know'. His second talk in August was entitled: 'The Sub Seafloor Ocean: A Voyage of Discovery based largely on Ocean Drilling'.

General outreach

Outreach activities remain central to our mission. In both 2012 and 2013, we funded 20 university undergraduate students to attend an ANZIC Marine Geoscience Masterclass in Perth, with the aim of inspiring the next generation of bright young scientists to work in this exciting area of research. Their feedback was very positive.

Partly in order to gauge the level of knowledge that academia, industry and government had regarding the IODP, the ANZIC Governing Council set up a cost–benefit analysis in late 2012, conducted under contract by the Allen Consulting Group and led by the highly experienced consultant Grahame Cook. The findings of this analysis are outlined in Chapter 12.

As the result of this study, in April 2013, the governing council decided to hire Grahame Cook to proceed with the initial stage of a long-term communication strategy designed to make government more aware of the activities and importance of the IODP. In the initial phase, he held very useful meetings with Dr Robert Porteous (Head of Research and Science Division, Department of Industry, Innovation, Climate Change, Science, Research and Tertiary Education (DIICCSTRE)) and Ms Patricia Kelly (Deputy Secretary, DIICCSTRE). In addition, by casting the net more broadly, constructive discussions were held, jointly with Neville Exon, with officers in the Climate Science and Adaptation Division of DIICCSTRE, Department of Foreign Affairs and Trade (DFAT) and Department of the Prime Minister and Cabinet. At DFAT's suggestion, contact was also made with officers in the Department of Sustainability, Environment, Water, Population and Communities (DSEWPaC).

These discussions confirmed that the level of awareness about IODP and its value to Australia among policy departments was quite low. Despite this, when departmental officers were briefed, there was a great deal of interest in and support for IODP. Officers in all the departments responded positively in seeking to understand the program better and its relevance to their ongoing and emerging areas of investigation. On this basis, the council decided to continue these activities in following years, with Grahame Cook and Neville Exon working as a team, with excellent results.

Port calls

There were port calls with associated ship visits and publicity in Townsville (November 2009), Wellington (January 2010), Hobart (March 2010), and Auckland (December 2010) by *JOIDES Resolution*, and in Townsville (February 2010) by *Greatship Maya*. All port calls drew numerous VIPs and generated favourable press coverage. The Hobart port call had the highest profile visitor, the Australian Minister for Innovation, Industry, Science and Research, the Honourable Kim Carr.

Townsville *JOIDES Resolution* port call before Canterbury Basin Expedition 317, November 2009

The *JOIDES Resolution* docked in Townsville from 4 to 6 November, with a press conference aboard the ship early on Friday 6 November. The ship was leaving Townsville for a two-month expedition to drill the Canterbury Basin east of New Zealand's South Island. This expedition investigated the relative importance of global climate change versus local tectonic forces on changing sea level and sedimentary processes during the last 30 million years. This sedimentary basin is a good place to investigate past global sea-level changes, which have frequently amounted to ~100 m. An understanding of past sea-level changes helps geologists to better interpret sedimentary strata around the world, which is important for resource assessment.

Neville Exon and Sarah Howgego from the ANZIC IODP Office in Canberra flew to Townsville for the occasion, and for the associated public relations exercises, including a number of ship tours. Jim O'Brien of James Cook University (JCU) was a marvellous help with public relations, and the various media releases came out from JCU. On Friday 6 November, we had a press conference and tour aboard the ship at which the speakers were David Divins of Ocean Leadership, introducing the ship; Neville Exon, introducing ANZIC; the local federal Member of Parliament, James Bidgood; Will Sager, Shatsky Rise Expedition 324 co-chief scientist, who was coming off the ship; Craig Fulthorpe, Canterbury Basin Expedition co-chief scientist; and Bob Carter of JCU and Simon George of Macquarie University, Australian Canterbury Basin expeditioners.

Other VIPs attending the event included Dr Ian Poiner, CEO of the Australian Institute of Marine Science (AIMS); Professor Bill Collins from JCU; and Dr Kate Wilson, Chair of the ANZIC Governing Council. All were very impressed with the ship and its program. There were about 15 print, radio and TV journalists aboard, who produced mostly local coverage of the event. Channel 9 filmed the activity, and there was online coverage from the ABC. It was a great opportunity for us to see the wonderful new science facilities installed on the ship in Singapore as part of a US$120 million refit.

On Thursday and Friday, another 100 or so invitees toured the ship, including AIMS and JCU staff, JCU students, and high school students from Calvary Christian College and The Cathedral School. Sarah Howgego had made the arrangements on land with JCU help, and Sarah Saunders (Ocean Leadership) organised the excellent tours aboard ship. David Murphy from Queensland University of Technology was a superb tour guide after taking part in the previous Shatsky Rise expedition.

Wellington *JOIDES Resolution* port call between Canterbury (317) and Wilkes Land (318) Expeditions, January 2010

GNS Science and Victoria University of Wellington hosted the Wellington port call of the *JOIDES Resolution* in January, between the Canterbury Basin and Wilkes Land expeditions. Around 100 VIPs, scientists, petroleum geologists and university students toured the ship and attended lunchtime and evening talks. Twenty high school students participated in a holiday program centred around the *JOIDES Resolution* visit that included a ship tour and collection and analysis of a gravity core from Wellington Harbour. All events proved very successful, had good media coverage (national TV and radio and, Wellington newspapers) and were well-attended by senior representatives from government agencies, research institutes and universities.

Townsville *Greatship Maya* port call before Great Barrier Reef Environmental Change Expedition 325, February 2010

In early February 2010, the 93 m long *Greatship Maya* (brand new from a Singapore dockyard) sailed from Townsville to drill short core holes in the deep-water terraces of the ancient Great Barrier Reef to study environmental changes in the last 30,000 years. This was an alternative platform expedition supported by ECORD. Jody Webster of the University of Sydney and formerly of James Cook University was one of the two co-chief scientists for this expedition. Other Australian researchers were, and some still are, participating intensively in the project.

In co-operation with ANZIC (Neville Exon and Sarah Howgego) and other Australian partners, the ECORD Outreach team, led by Albert Gerdes, organised a number of events. In the months ahead of the expedition, an extensive list of Australian media outlets was compiled. Journalists and local VIPs were then invited to a media conference and media releases were distributed. On 11 February, a very well attended news conference was held in the Jupiters Hotel. Among others, the co-chief scientists and the expedition project manager informed the journalists about the science and logistics of the expedition. The conference was completed with a tour of the vessel, during which the journalists had the opportunity to interview a number of expedition participants. The feedback from the journalists was that they were very satisfied with the event.

In the days before the media conference, both the co-chief scientists and representatives of ANZIC and ECORD had already given interviews for radio and print media. Along with the media conference, a press release was distributed through the worldwide IODP channels. The Australian media response included reports about the expedition that were published by ABC TV and Radio, the *Cairns Post* and other privately owned television and radio stations. The media release was also picked up by international online and print media in Japan, India, Europe and the Fiji Islands. The ECORD Outreach team produced two posters and an audiovisual show that were exhibited at the Reef HQ Aquarium in Townsville.

A second media conference was organised during the onshore science party meeting that was held in July 2010 in the IODP Bremen Core Repository. The event was attended by regional TV teams, radio journalists, photographers and journalists from regional print media and the national German news agencies respectively. Again, a media release was distributed on the same day providing information about the findings of the science party. Articles appeared in 20 German newspapers.

The co-chief scientists starred in an ABC nationwide science program and gave an interview for the Queensland branch of ABC; a few international media outlets also picked up the news. Last, but not least, both the start of the expedition and the results achieved during the science party were communicated within the IODP community.

Overall, the media campaign for Expedition 325 was one of the most successful for all the mission-specific platform operations carried out by ECORD.

Figure 8.1. The media filming *Greatship Maya* in Townsville before Expedition 325

Source: A. Gerdes of ECORD

Hobart *JOIDES Resolution* port call after Wilkes Land Expedition 318, March 2010

A major outreach and publicity effort was made after the Wilkes Land Expedition 318, which was investigating the changes in climate and oceanography in the last 55 million years, as Australia and Antarctica separated, Antarctica froze over and global climate cooled. The *JOIDES Resolution* port call featured three days of ship tours and a well-attended public lecture at the University of Tasmania by expedition co-chief scientist Henk Brinkhuis. Both the University of Tasmania and IODP media groups were involved and this resulted in excellent TV, radio and press coverage in Tasmania and beyond. The organising group included Neville Exon and Sarah Howgego from the ANZIC IODP Office, Brad Clement and two others from the IODP managing group in Texas, and David Divins and Kristin Ludwig from the Ocean Leadership publicity group. Dr Kevin Welsh, an expeditioner from the University of Queensland, was among the tour guides.

On Wednesday 10 March, about 60 people toured. This included 30 scientists from the University of Tasmania, including Professor David Green, CSIRO, including Dr Ian Cresswell, and ANU, including Professor Andrew Roberts. We also showed 30 students from the University of Tasmania over the ship. We were particularly honoured to host the Governor of Tasmania, His Excellency Peter Underwood, who gave us a much-appreciated reception at Government House later that day.

Thursday was given over to tours for 65 senior school students interested in science, and their teachers. They came from Ogilvie High School, New Town High School, Rose Bay High School and Kingston High School. They were very interested and asked intelligent questions about the ship and the science.

Friday was devoted to a VIP and media tour, with Senator Kim Carr leading the group. He spoke of the importance of marine science to Australia, linked the newly planned Australian research vessel *Investigator* to IODP research, and was given some Miocene sediment as a memento. Among the other dozen eminent guests were Professor Daryl Le Grew, Vice-Chancellor of the University of Tasmania; Dr Bruce Mapstone, Chief of the CSIRO Division of Marine and Atmospheric Research; Dr Geoff Garrett, former Chief of CSIRO and soon-to-be Chairman of the ANZIC Governing Council; Professor Michael Stoddart, Director of the Institute of Marine and Antarctic Studies; Dr Tim Moltmann, Director of the Integrated Marine Observing System; and Professor Ross Large, Director of the Centre of Excellence in Ore Deposits.

All in all, this was a hugely successful exercise in bringing IODP science to the attention of the Minister for Science and Research, the Tasmanian scientific community and the media.

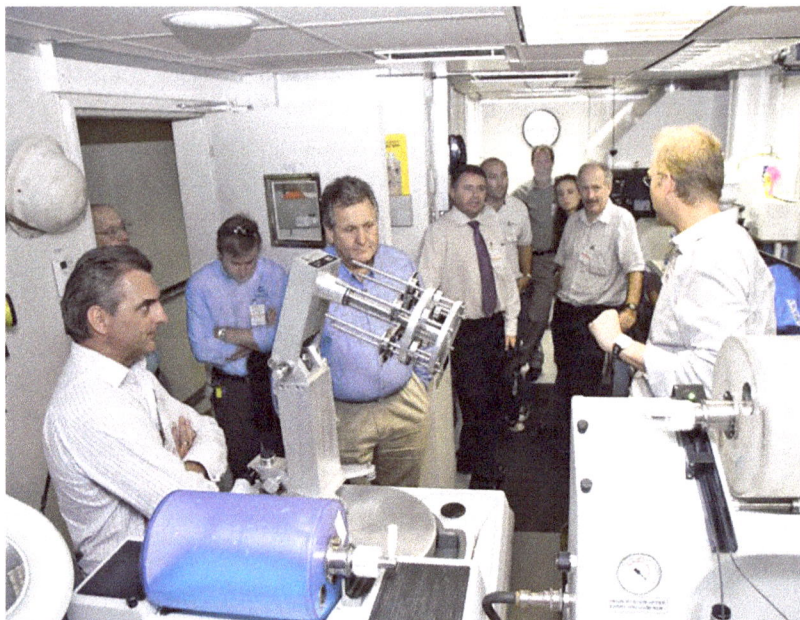

Figure 8.2. Australian dignitaries aboard *JOIDES Resolution* in Hobart
Source: Neville Exon

Auckland *JOIDES Resolution* port call between South Pacific Gyre Microbiology (329) and Louisville Seamount Chain (330) Expeditions, December 2010

This port call in Auckland, on 14 and 15 December, was organised by GNS Science (Giuseppe Cortese), the Auckland Museum Institute (Andrea Webley) and the University of Auckland (Lorna Strachan), with support from IODP and Ocean Leadership (Sarah Saunders). The activities included eight ship tours (112 people had a detailed visit of the ship), a guided tour of the Auckland Museum with a VIP lunch and reception (65 people registered) for the expedition scientists, and several talks about the bracketing expeditions (South Pacific Gyre Microbiology and Louisville Seamount Chain). The VIP reception featured talks by Mitch Malone (IODP) and by co-chief scientists Steven D'Hondt (University of Rhode Island, South Pacific Gyre Microbiology Expedition) and Anthony Koppers (Oregon State University, Louisville Seamount Trail Expedition).

An evening lecture for the public was also held, with 70 people registering. It included talks by Bruce Hayward (Geomarine Research), Christian Ohneiser (University of Otago), David Smith (University of Rhode Island) and Anthony Koppers (Oregon State University).

A meeting was also arranged at the Auckland Museum with representatives from the Ministry of Science and Innovation and various leading scientific figures from several New Zealand research institutes and universities, in order to stress the importance of scientific drilling for New Zealand. The aim was to strengthen the case for a stable funding base for New Zealand's IODP membership.

These activities generated a lot of media interest including from Radio NZ, TV3 New Zealand, the *New Zealand Herald*, the NZ Press agency, and Australia ABC Science. Steven D'Hondt (University of Rhode Island, USA, co-chief scientist South Pacific Gyre Expedition) and John Moreau (University of Melbourne, shore-based scientist) were interviewed by the Australian Science Media Centre about their interest in discovering, quantifying and understanding microbial life forms adapted to live in a large area of the Pacific Ocean known to be very poor in nutrients. A news item reporting on the *JOIDES Resolution* Auckland port calls was published in the journal *Scientific Drilling* (No. 11, March 2011, p. 76).

9

Major IODP planning workshops

Neville Exon

Planning workshops for proposed future IODP drilling are critically important in generating ideas that can lead to successful drilling proposals that can address global scientific questions. Among other things, they bring together a critical mass of scientists to be able to carry out informed discussions, and often to form the core of proponents of future proposals. Four workshops of special interest to Australia and New Zealand are discussed below.

Such workshops can be global in scope, such as the IODP New Ventures in Exploring Scientific Targets (INVEST) workshop in 2009, which addressed the whole question of the way ahead for scientific ocean drilling, and led to the preparation of the carefully thought through, and hugely influential, IODP Science Plan for 2013–2023. Another global workshop was the Chikyu+10 workshop in 2013, designed to help plan IODP scientific ocean drilling over the next decade using the unique deep-drilling riser capacity of the *Chikyu*. Implementation of the workshop's plans has been hampered by funding limitations.

The next level of IODP workshops is those covering a large region, bringing together a varied group of scientists interested both in that region and in developing proposals for future IODP drilling. Such workshops were the Indian Ocean IODP Workshop in 2011 and the Southwest Pacific IODP

Workshop in 2012. Both generated a great deal of interest and helped lead to numerous successful IODP proposals, which were drilled from 2014 to 2016 and will be drilled in 2017 and 2018.

Finally, there are more detailed planning workshops dealing with a particular topic or even with a single proposal, but none are discussed below.

INVEST meeting, Bremen, September 2009

In September 2009, a very large (584 participants) scientific meeting was held in Bremen, Germany, to begin designing the next 10-year phase of IODP to start in late 2013. Several Australians played key roles at this meeting and afterwards. Australians attending were Brian Kennett, Richard Arculus, Stephen Gallagher, Jody Webster, Chris Yeats, Helen McGregor and Elizabeth Abbey. The INVEST meeting report is available from usoceandiscovery.org/past-workshops-old/iodp-new-ventures-in-exploring-scientific-targets-invest.

Preceding the meeting, the Science Committee of ANZIC had produced a white paper entitled *IODP INVEST: Beyond 2013, Australasian White Paper*, authored by Stephen Gallagher of the University of Melbourne and others (usoceandiscovery.org/wp-content/uploads/2016/05/INVEST_Report.pdf). Its initial statement was the following:

> In order for international collaborative ocean drilling to continue as a major research effort post-2013, it is necessary for the scientific community to demonstrate the societal relevance and impact of their work on a global scale. Increasingly, national earth science funding around the globe is focusing on research that directly influences society, most notably in the areas of climate change and hazard mitigation. Australia and New Zealand are unique among the members of IODP in that they are the only countries in the Southern Hemisphere. They direct scientific access to, and have direct national interest in, a vast region that extends from the Equator to Antarctica. This region incorporates the climatologically vital Southern Ocean and the active tectonic margins of the Southwest Pacific and Indian Oceans, major global sources of tsunami activity. Furthermore, the Australian and New Zealand earth science community includes world leaders in their fields who possess unique knowledge and expertise critical to any global scientific investigation. Bearing this in mind, we believe that our two countries can make a major contribution to the INVEST planning process and the ocean drilling research effort moving forward.

An ANZIC report on the meeting was written by Stephen Gallagher, Chris Yeats, Chris Hollis, Jody Webster and Helen McGregor, but was never published. In summary they wrote:

> Over 570 scientists from around the world attended a series of workshops hosted by MARUM at Bremen, 23–25 September 2009. The organising committee allocated each delegate to 2 to 3 working group (WG) sessions over 3 days, which were facilitated by a WG leader and scribe. Outcomes of each workshop were communicated to all delegates in several joint sessions and as a series of reports to the organising committee. It is clear that there are many unanswered questions that need to be addressed by the IODP beyond 2013. However, it will be an onerous task to distil all the contributions into a science plan that is sufficiently different from the Initial Science Plan to convince the various funding bodies to continue to fund IODP beyond 2013.

Helen McGregor made some general observations:

> As outlined below numerous exciting hypotheses were put forward under each theme for the IODP post-2013 period. Several of the Working Group summaries advocated new, though similar approaches to conducting IODP science, and testing their respective hypotheses.

> The common approaches included in her observations were:

> **Transects**. For example, shelf to basin transects would link terrestrial and marine climate variability, and/or evaluate the timing of marine and terrestrial biological response to changing climates. Latitudinal transects were suggested to investigate tropical ocean–atmosphere dynamics, and pole–equator climate and sea-level variability, and to improve spatial coverage across key periods e.g. Palaeocene-Eocene Thermal Maximum (PETM) and Oceanic Anoxic Events (OAEs) (climate and ocean acidification questions). Other suggestions included ridge to passive/active margins transects to understand the evolution of ocean crust through time, and transects across subduction zones.

> **Recovery**. Many groups called for continuous records, which would allow them to better test their hypotheses. One example was improved recovery through past warm periods. Higher recovery also would help to better investigate subduction zones and volcanic arcs processes.

> **High-resolution**. Obviously the drill core resolution required is dependent on the question being asked but higher-resolution records were requested for studying past variability in interannual climate modes such as ENSO, shoreline dynamics in response to sea-level changes, higher resolution 'windows' across periods such as the PETM, and the evolution of sea-floor

biota. Higher-resolution records were also requested for investigating past ecosystem responses to critical events in Earth's history, and the behaviour of the geodynamo.

Observatories. The use of real-time observatories was proposed to better constrain biogeochemical cycles. In situ measurement and monitoring of seismic activity were proposed to understand the triggers of earthquakes, submarine landslides and tsunamis. Better linkage of ocean climate records with instrumental data was also proposed.

Data–model comparisons. This was clearly advocated within the Climate Change theme. Input from the modelling community would improve understanding of volcanic geohazards and ocean acidification.

Richard Arculus of The Australian National University (ANU) was a member of the post-INVEST New Science Plan Writing Committee consisting of 14 scientists, as was New Zealander Peter Barrett of Victoria University, Wellington, showing the scientific regard in which ANZIC is held, with no other Associate Member country represented on this committee. The workshop and the writing committee were the genesis of the exciting IODP Science Plan for 2013–2023, *Illuminating Earth's Past, Present, and Future*, available from www.iodp.org/about-iodp/iodp-science-plan-2013-2023.

Chikyu+10 Workshop, Tokyo, April 2013

A three-day Chikyu+10 Workshop was held in Tokyo to help plan IODP scientific ocean drilling over the next decade using the unique deep-drilling riser capacity of the *Chikyu*. It was chaired by Mike Coffin from the University of Tasmania and 397 participants attended from 21 countries, including 15 Australians and New Zealanders, who were largely funded by ANZIC. The workshop had five themes that reflected IODP's new Science Plan: dynamic fault behaviour, continent formation, deep life, ocean crust and Earth's mantle, and sediment secrets. Full workshop proceedings are available from www.jamstec.go.jp/chikyu+10/docs/C+10_report_textbody.pdf.

It was apparent that *Chikyu*'s riser capacity would be fully utilised until 2017 in completing the earthquake-related Nankai Trough subduction zone NanTroSEIZE program, with the plan being to drill 5,000 m below the sea floor. At that time, the choices for a succeeding IODP deep-drilling program, with a broader geographical spread, included drilling:

- the CRISP subduction zone program off Central America
- the Izu-Bonin Arc south of Japan
- the Hikurangi margin subduction zone east of New Zealand (IODP Proposal 781B)
- a Mohole program to reach the Earth's mantle, perhaps off Hawaii.

Much detailed work is going into these potential programs. *Chikyu* is also seeking commercial work, and would consider doing IODP non-riser drilling if this made logistical sense.

More recently, IODP Full Proposal 871(CPP) has been submitted, entitled 'First deep stratigraphic record for the Cretaceous eastern Gondwana margin: Tectonics, palaeoclimate and deep life on the Lord Howe Rise high-latitude continental ribbon'. It would use *Chikyu* under a joint agreement between Geoscience Australia and JAMSTEC. The main aim of the expedition would be to better understand the geological history of this part of the former Australian margin, but the cores would also be critical to understanding its petroleum potential. Things have progressed to the extent that drilling might occur this decade.

Indian Ocean IODP Workshop in Goa, India, October 2011

This workshop was initiated and planned by ANZIC and IODP-India, on the understanding that *JOIDES Resolution* could be encouraged to visit the Indian Ocean in 2014 for the first time in a decade. It was hosted by the National Centre for Antarctic and Ocean Research (NCAOR), with Dhananjai Pandey as coordinator. The workshop aimed to improve existing proposals, build new proposals and familiarise our Indian colleagues with IODP, which India had joined not long beforehand. Neville Exon, Stephen Gallagher, Mike Coffin and Richard Arculus played leading roles, and there were another eight ANZIC participants. This was a very high profile event in India, and about 40 foreigners and 70 Indian scientists attended the workshop.

Figure 9.1. Dignitaries at the opening ceremony, with Dhananjai Pandey (co-convenor from NCAOR) at the lectern and Neville Exon (co-convenor from ANZIC) beside him

Source: Indian National Centre for Antarctic and Ocean Research

Figure 9.2. Participants in the Goa workshop

Source: Indian National Centre for Antarctic and Ocean Research

The workshop themes were:

- **Cenozoic oceanography, climate change, gateways and reef development**. This covered both very broad questions related to the Indian Ocean, and narrower ones such as the causes and effects of the Indonesian Throughflow Current, sea-level rise and fall, and the origin of late Pleistocene reefs.
- **The history of the monsoons**. This covered tectonics, uplift, weathering and erosion, sediment deposition, and climate and oceanography, and dealt with all Indian Ocean areas affected by monsoons.
- **Tectonics and volcanism**. There are many questions related to the tectonism of the Indian Ocean, such as plate tectonics, the evolution of the oceanic crust including mid-ocean ridge formation and the formation of large igneous provinces, continental rifting and related deposition, subduction, and arc volcanism and earthquakes.
- **The deep biosphere**. Pioneering studies of the extremophiles of the deep biosphere in sediments and basalts have largely been concentrated in the Atlantic and Pacific Oceans. Given the different nature of the oceanography and inputs of organic matter into the Indian Ocean, the deep biota could be rather different there.

Many papers were presented and a full summary of the workshop, entitled 'Detailed Report on International Workshop on Scientific Drilling in the Indian Ocean, Goa, India, October 17–18, 2011', is available from usoceandiscovery.org/wp-content/uploads/2016/05/Workshop_Report_India.pdf. There was much discussion of existing and future proposals, and ANZIC involvement in a variety of proposals was expected. The aim was to have proposals submitted for the 2012 meetings of the IODP Proposal Evaluation Panel, with the hope that they would be drilled by *JOIDES Resolution* in 2014, or later by alternative platforms. The existing Northwest Shelf sea level proposal, in the Dampier Basin, was discussed and was to be simplified to cover the last 5 million years, including a strong component dealing with the Indonesian Throughflow Current. Other proposals already in the mix in our region included drilling on the Naturaliste Plateau (Cretaceous black shales), Kerguelen Plateau (Cretaceous Large Igneous Province) and Great Australian Bight (biogenic gas-rich Quaternary sediments). The pirate threat in the Arabian Sea meant that there was a shortage of feasible highly ranked proposals in the Indian Ocean, so there was a great need for excellent new proposals outside the Arabian Sea.

This workshop was an important step toward more IODP drilling in the Indian Ocean, and increasing international cooperation for that and other geoscientific purposes. It has borne fruit with the first ocean drilling in the Indian Ocean for many years. Remarkably, eight Indian Ocean IODP expeditions were later approved for 2015 and 2016 and drilled (see Chapter 10, Figure 10.1).

The workshop has been covered in a short *Eos* article (Vol. 93, No. 7, 14 February 2012), in the full report cited above, and was later in a report in *Scientific Drilling* No. 14 (September 2012, pp. 60–67; usoceandiscovery.org/wp-content/uploads/2016/05/Workshop_India_SD_Report.pdf).

Southwest Pacific IODP Workshop in Sydney, October 2012

Figure 9.3. Co-convenor Stephen Gallagher addressing the workshop
Source: Neville Exon

With a diverse group of 80 scientists from around the world, this workshop was held at the University of Sydney, in order to review the latest research in the region, briefly outline possible future IODP expeditions and set up working groups to develop compelling new drilling proposals in the global science context. As the *JOIDES Resolution* was expected to be in the region fairly soon, the workshop participants agreed on the urgent need to build strong science proposals. The workshop was hosted by ANZIC and the University of Sydney, with additional funding from IODP-Management International, the US Science Support Program and Japan Drilling Earth Science Consortium. It covered all fields of geoscience, and drilling targets that extended from the Equator to Antarctica. The four science themes of the new IODP Science Plan were addressed. An additional resource-oriented theme considered possible co-investment opportunities involving IODP vessels.

Various new full and add-on proposals were identified, with the aim of submitting most by the proposal deadline in April 2013:

- **Climate and Ocean Change**: marine Palaeogene proposals, namely Lord Howe Rise and Campbell Plateau, and a Wilkes Land continental shelf Neogene proposal.

- **Deep Biosphere**: biosphere in organic-rich Gulf of Papua sediments, and several ancillary proposals.

- **Earth Connections**: formation of the Greater Ontong Java large igneous province, initiation of subduction and origin of sedimentary basins in the Lord Howe Rise region, and structure and dynamics of mantle flow in the northern Australian-Antarctic Discordance.

- **Earth in Motion**: the active Brothers Volcano system in the Kermadec Arc, active volcanic systems in the Manus Basin, the nature of the Tuaheni Landslides off northeast New Zealand, and near-trench-axis comparative drilling around the Pacific Ocean.

- **Marine Resources**: the nature and resource potential of gas hydrates off northeast New Zealand, and deep stratigraphic drilling on the Lord Howe Rise related to both petroleum potential and research.

Many of the proposals were broad and multidisciplinary in nature, hence optimising the scientific knowledge that can be produced by the use of IODP infrastructure. This was particularly the case for the proposals related to active volcanic systems in the Brothers Volcano and Manus Basin; the Cretaceous-Palaeocene palaeoenvironment, tectonic history, and petroleum potential of the Lord Howe Rise region; and slow-slip subduction, fluid flow, landslides and gas hydrate potential of the Hikurangi subduction margin.

Many drilling ideas were put forward for consideration, many of them were later submitted as proposals, and most of the proposals are now approved (see Chapter 10, Figure 10.1). Several lead proponents, who are now co-chief scientists, have remarked that without the workshop they never would have written proposals.

The workshop results were published in detail on the IODP website in early 2013 (usoceandiscovery.org/wp-content/uploads/2016/05/Workshop_SWPacific_Report.pdf), and later in *Scientific Drilling* No. 17 (April 2014, pp. 45–50).

10

The future of scientific ocean drilling: International Ocean Discovery Program

Neville Exon

Various national and international reviews of ocean drilling were held in recent times, some of them focused tightly on the Integrated Ocean Drilling Program itself, while other reviews also focused on its predecessors. This was obviously necessary when considering renewal of a program that costs about US$180 million annually for its logistics, and has two large drill ships with a replacement value of about US$1.1 billion. The additional costs of the science participants (carried by their own countries) amount to many millions of dollars.

A 10-year phase of ocean drilling from 2013 to 2023 was approved under the new name International Ocean Discovery Program (IODP(2)). The change of name from Integrated Ocean Drilling Program (IODP(1)) was justified because of the broadening of the activities to include suites of borehole observatories for which drilling is just an enabling tool. Key funding decisions determining the future scope of IODP were made by the US National Science Foundation (NSF), the Japanese Ministry of Education, Culture, Sports, Science and Technology (MEXT) and the European Consortium for Ocean Research Drilling (ECORD) in late 2013. The structure of the new program is much looser than the previous one, with those who provide the vessels – the US, Japan and Europe –

having ultimate control of their programs. Australian and New Zealand scientists have helped design proposals for funding for the new IODP, with some already carried out.

The latest IODP Science Plan *Illuminating Earth's Past, Present, and Future* was published in early 2013 (www.iodp.org/science-plan/127-low-resolution-pdf-version/file) and summarised in www.iodp.org/science-plan/115-iodp-science-plan-br/file. Its scientific themes are:

Climate and Ocean Change: Reading the Past, Informing the Future
Ocean floor sediment cores provide records of past environmental and climatic conditions that are essential for understanding Earth system processes.

Biosphere Frontiers: Deep Life and Environmental Forcing of Evolution
Samples recovered by ocean drilling permit study of Earth's largest ecosystems, offering insight into the origins and limits of the deep biosphere, the evolution of marine microfauna through times of environmental change and human evolution related to climate change.

Earth Connections: Deep Processes and their Impact on Earth's Surface Environment
The dynamic processes that create and destroy ocean basins, shift the position of continents and generate volcanoes and earthquakes extend from Earth's core to its atmosphere, and are fundamental for understanding global change within the context of planetary evolution.

Earth in Motion: Processes and Hazards on Human Time Scales
Many fundamental Earth system processes, including those underlying major geologic hazards, occur at 'human' time scales of seconds to years, requiring new sampling, downhole measurement, monitoring and active experimental approaches.

Australia and New Zealand in IODP: 2014 and 2015

Australia and New Zealand remain enthusiastic supporters of ocean drilling. We joined the new phase of IODP(2) in September 2013 as the Australian and New Zealand IODP Consortium (ANZIC). Australia was funded for 2014 and 2015 under an ARC/LIEF grant, with agreed funding from our partners. The New Zealand partners funded themselves, with GNS Science paying the lion's share of their costs.

Australian IODP partners

The Australian National University
CSIRO Earth Science and Resource Engineering
Curtin University of Technology
Geoscience Australia
James Cook University
Macquarie University
Monash University
MARGO (Marine Geoscience Office)
Queensland University of Technology
University of Adelaide
University of Melbourne
University of New England
University of Queensland
University of Sydney
University of Tasmania
University of Technology, Sydney
University of Western Australia
University of Wollongong

New Zealand IODP partners

GNS Science
NIWA
University of Otago
Victoria University of Wellington

Australia and New Zealand in IODP: 2016 to 2020

Australia and New Zealand will continue as ANZIC in IODP. Australia has been well funded for 2016 to 2020 under another ARC/LIEF grant. Unfortunately, James Cook University, the University of New England and the University of Technology of Sydney have withdrawn, but the University of New South Wales has joined. Total annual funding from ARC and our Australian partners has increased somewhat. The New Zealand partners will fund themselves and the University of Auckland has joined their consortium. The long-term nature of the new funding is very welcome, allowing us to plan for the longer term at a time when there will be much IODP activity in our region (see Figure 10.1 below).

Australian IODP partners

The Australian National University
CSIRO Earth Science and Resource Engineering
Curtin University of Technology
Geoscience Australia
Macquarie University
Monash University
MARGO (Marine Geoscience Office)
Queensland University of Technology
University of Adelaide
University of Melbourne
University of Queensland
University of Sydney
University of Tasmania
University of New South Wales
University of Western Australia
University of Wollongong

New Zealand IODP partners

GNS Science
NIWA
University of Auckland
University of Otago
Victoria University of Wellington

IODP Regional Expeditions: 2014 to 2018

Three new expeditions were carried out in our region in 2015 and 2016: Indonesian Throughflow Expedition 356, August–September 2015; Sumatra Seismogenic Zone Expedition 362, August–September 2016; and Western Pacific Warm Pool Expedition 363, October–November 2016. Five more regional expeditions are approved in 2017 and 2018 as shown in Figure 10.1 below. Furthermore, it should be noted that nearly 9 per cent of active but not approved proposals are led by ANZIC scientists (Figure 10.2), and that 53 per cent of all active proposals are in the Indian and Pacific Oceans (Figure 10.3), so it is very likely that there will be more drilling in our region before 2023, when the present phase of IODP ends. To help ensure that drilling will happen, an Australasian IODP Regional Planning Workshop was held in Sydney in June 2017.

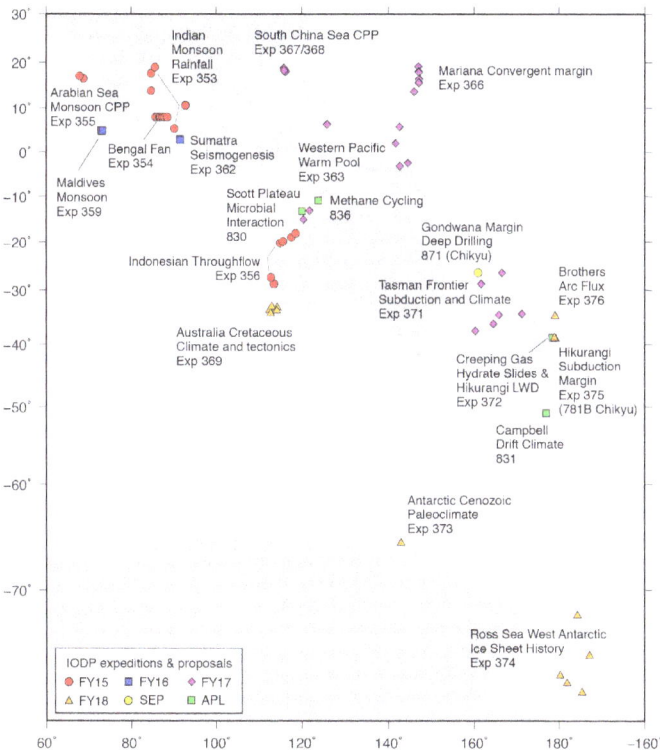

Figure 10.1. Completed, proposed and approved IODP Expeditions in the Australasian region: 2015–2018

SEP means with the IODP Science Evaluation Panel; APL means a short Ancillary Project Letter.

Source: Katerina Petronotis of *JOIDES Resolution* Science Operations, map dated May 2016

All these expeditions have used or will use *JOIDES Resolution*, except for Expedition 373 and Proposals 781B and 871. Note that the US fiscal year is used in these figures; for example, FY17 began in October 2016.

Active proposals: 92
by lead proponent's member affiliation

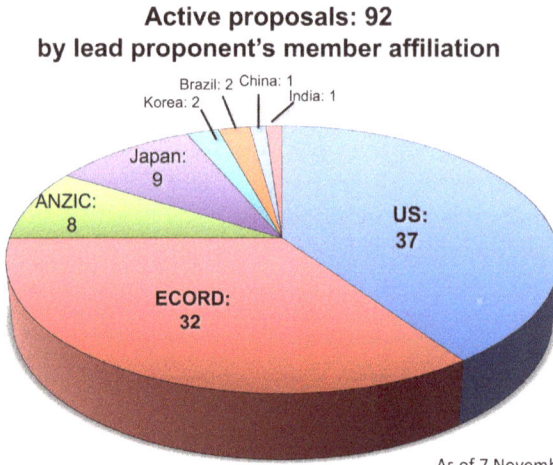

As of 7 November 2016

Figure 10.2. Active IODP proposals by lead proponent's member affiliation
Source: IODP Science Support Office at Scripps Institution of Oceanography

Active proposal status: 92

by target ocean

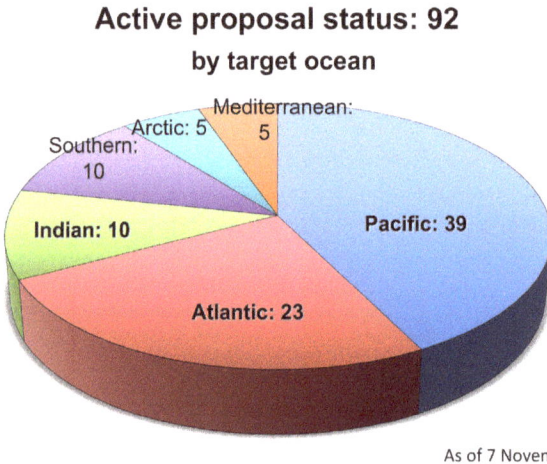

As of 7 November 2016

Figure 10.3. Active IODP proposals by target ocean
Source: IODP Science Support Office at Scripps Institution of Oceanography

11

Regional proposals drilled since 2013 and to be drilled soon

The quality of drilling proposals for the period after 2013, often led by ANZIC scientists, is very high, as is attested by the articles below. They were all written in 2016. The expeditions have all been drilled or are scheduled to be drilled. Proposal 832 has become Expedition 371; Proposal 781A will become Expeditions 372 and 375; Proposal 751 will become Expedition 374; and Proposal 818 will become Expedition 376. ANZIC authors, other than the writers of the articles, are shown in bold print.

>>

Indonesian Throughflow: A 5-million-year history of the Indonesian Throughflow Current, the Australian Monsoon and subsidence on the Northwest Shelf of Australia: IODP Expedition 356

Stephen J. Gallagher, University of Melbourne

(On behalf of Craig S. Fulthorpe, University of Texas at Austin; Kara Bogus, *JOIDES Resolution* Science Operator, College Station, Texas; and the Expedition 356 Scientists)

Introduction

In 2015, IODP Expedition 356 successfully addressed three important topics off northwest Australia in the last 5 million years: the history of the major currents, the onset of the monsoon and regional aridity, and the regional subsidence history. The Indonesian Throughflow Current is a critical part of the global oceanic thermohaline conveyor. It transports heat from the equatorial Pacific (the Indo-Pacific Warm Pool) to the Indian Ocean, thus exerting a major control on global climate. The complex tectonic history of the Indonesian archipelago, a result of continued northward motion and collision of the Australasian plate into the southeast Asian part of the Eurasian plate, makes it difficult to reconstruct long-term (i.e. million-year) throughflow history from sites within the archipelago. The best areas to investigate Indonesian Throughflow history are downstream in the Indian Ocean, either in the deep ocean away from tectonic deformation or along passive (non-tectonic) margins directly under its influence. Although previous ODP and DSDP deep-water cores in the Indian Ocean have been used to document the variability of the Indo-Pacific Warm Pool and the Indonesian Throughflow, these sections lack direct evidence of the current. IODP Expedition 356 cored seven sites covering a latitudinal range of 29°S–18°S off the northwest coast of Australia to obtain a 5-million-year record of Indonesian Throughflow, Indo-Pacific Warm Pool, and climate evolution that matches deep-sea records in its resolution.

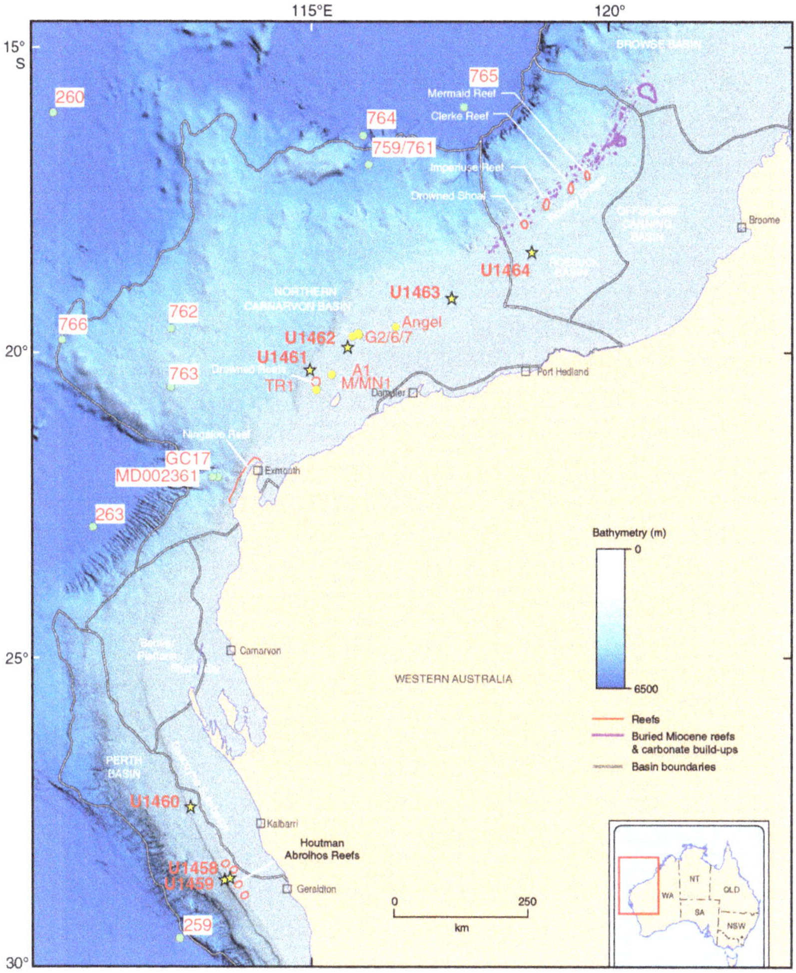

Figure 11.1. Map of the Northwest Shelf showing major basins and location of modern and 'fossil' reefs

Stars = drill sites; green circles = DSDP/ODP sites and other core locations referred to in text; yellow circles = industry well locations (Angel = Angel-1; G2/6/7 = Goodwyn-2, Goodwyn-6, and Goodwyn-7; A1 = Austin-1; M/MN1 = Maitland/Maitland North-1; TR1 = West Tryal Rocks-1). WA = Western Australia, NT = Northern Territory, SA = South Australia, QLD = Queensland, NSW = New South Wales.

Source: Stephen Gallagher

The history of the Australian Monsoon and its variability are thought to be related to the East Asian Monsoon, and this is hypothesised to have been in place since 5 million years ago (Ma). The presence of airborne desert dust and the pollen of fossil plants in our cores better constrains the development of aridity on the Australian continent. Detailed palaeobathymetric and stratigraphic data from the well transect address the subsidence history over the last 5 million years, and its causes.

Summary of outcomes

Expedition 356 began in Fremantle, Western Australia, on 31 July 2015. After a short transit, coring began near 29°S and progressed northward to 18°S along the Northwest Shelf of Australia (Figure 11.1). Seven sites recovered 5,185 m of material with 67 per cent recovery overall. This is an unparalleled sediment archive of warm temperate to tropical climate and oceanography along the continental margin of Australia. Post-cruise research is providing insights into the complex relationship between subsidence, reef development, Indonesian Throughflow variability, the Australian monsoon and the onset of aridity in Australia over the last 5 million years. Despite intermittent recovery in some sections, the cores provide sufficient high-quality material for future studies by the Expedition 356 scientific party as well as many other scientists over the coming decades. Shipboard findings met or exceeded expectations, and we successfully addressed the three expedition objectives as summarised below.

1. Determine the timing and variability of the Indonesian Throughflow and the Indo-Pacific Warm Pool, and the onset of the Leeuwin Current, to understand the controls on Quaternary tropical carbonate and reef deposition.

Sites U1458, U1459 and U1460 (Figure 11.1), in the Perth Basin at ~29°S, are close to the Houtman-Abrolhos reef system, the most southerly reef system in the Indian Ocean. The high-latitude position of these reefs is related to the path of the south-flowing warm Leeuwin Current, which is itself controlled by the relative intensity of the Indonesian Throughflow. Hard cemented layers at the seabed prevented Site U1458 from successful coring, so we quickly proceeded to Site U1459. Sites U1459 and U1460 yielded 50- to 5-million-year-old carbonate-rich strata that record the role of the Leeuwin Current on the initiation and evolution of reefs on the Rottnest Shelf and the Carnarvon ramp. Sites U1461 and U1462 in the northern Carnarvon Basin are near a series of drowned reefs at ~22°S. From cored layers equivalent to those beneath these reefs, we can confirm that these are relatively young reefs that were initiated in the last million years. Site U1464 in the Roebuck Basin is close to a reef, and coring reveals that this Rowley Shoal drowned at least 3 Ma.

2. Obtain an approximately 5-million-year orbital-scale tropical to subtropical climate and ocean archive, directly comparable to deep-ocean oxygen isotope and ice-core archives, to chart the variability of the Australian monsoon and the onset of aridity in northwestern Australia.

There are no well-constrained orbital-scale climate records older than 500,000 years along the entire western margin of Australia. We continuously cored strata to extend this record to 5 million years. In particular, we targeted areas that should yield significant dust, clay and pollen that have been transported from coastal regions off western Australia by wind or water (rainfall/monsoon). For example, Sites U1458–U1460 cored records of the southern Australian winter-dominated rainfall regime for more than the last 5 million years. The more northerly sites (U1461–U1464) cored records of the summer rainfall–dominated Australian monsoon for the same time period.

Post-cruise research on these sections is greatly increasing our understanding of how Australia's subtropical to warm temperate climate responded to Pliocene (5–2.5 Ma) warmth through to Pleistocene (2.5–0 Ma) icehouse conditions.

Site U1461 yielded the thickest (1,000 m) sequence of 5–0-million-year strata obtained during this expedition. Together with its excellent recovery (83 per cent) and the very good preservation of the microfossils, this is now one of the best sampled palaeoceanographic and climate archives along the western Australian continental margin. Calcareous microfossil preservation was also excellent at Site U1463. The very good preservation of microfossils at these sites will facilitate fine-scale palaeoceanographic and palaeoclimate reconstructions using inorganic and organic oceanic proxies leading to an unparalleled 5-million-year record of conditions downstream of the Indonesian Throughflow.

3. Provide empirical input into the patterns of subsidence along the Northwest Shelf that can be used to place fundamental constraints on the interaction between Australian plate motion and mantle convection.

Accurate subsidence analyses of the sections cored over 10° of latitude can resolve whether northern Australia moved with or over a time-transient or long-term stationary downwelling within the mantle, vastly improving our understanding of deep-earth dynamics and their impact on surface processes. The latitudinal transect of seven sites from 29°S to 18°S has provided a well-constrained framework that will allow construction of a series of detailed subsidence curves. Such analysis draws on the detailed rock physical properties data, age models and palaeodepth estimates from microfossils and seismic sections at each site.

Preliminary analyses of the four northerly sites have revealed that the Northwest Shelf subsided synchronously at around 5 Ma from palaeodepths of 0–50 m to over 300 m, and that subsequently this subsidence reversed. These results suggest that this part of the Australian plate travelled with a time-transient downwelling feature in the mantle. However, it is not clear how or why this process reversed. Further investigations of this unusual subsidence pattern will shed light on the impact of deep-earth dynamics (dynamic topography) on surficial processes.

Figure 11.2. The Assistant Minister for Science, the Honourable Karen Andrews MP, shares a joke with Stephen Gallagher, Brad Clement and Neville Exon beside *JOIDES Resolution* in Darwin after the expedition

Source: *JOIDES Resolution* Science Operator, Texas A&M University

The rise and fall of the Cretaceous hot greenhouse off southwest Australia: IODP Expedition 369

Richard Hobbs, Durham University

(On behalf of **Irina Borissova**, Geoscience Australia; Brian Huber, Smithsonian Institution; Kenneth MacLeod, University of Missouri)

Expedition objectives

Between 2015 and 2018, the drill ship *JOIDES Resolution* is undertaking several exciting expeditions in Australian waters. The expedition planned for September 2017 off southwest Australia will drill several holes on the Naturaliste Plateau and the adjacent Mentelle Basin lying beneath the Naturaliste Saddle. The unique tectonic and palaeoceanographic setting of this region offers an outstanding opportunity to investigate a range of scientific issues of global importance with particular relevance to climate change, as well as the geological history of western Australia.

Sites were chosen with the goal of recovering samples suitable for generating palaeotemperature and biotic records spanning the Cretaceous hot greenhouse, from initiation to collapse, including the times of Oceanic Anoxic Events (OAE) 1d (~98–97 Ma) and 2 (~94–93 Ma). Data generated will provide insight into surface and deep-water circulation that can be used to test predictions from and provide input for earth system models. The high palaeolatitude (60–62°S) location of the sites is especially important because high latitudes are more sensitive to climatic changes than mid and low latitudes. Such enhanced sensitivity should increase the size of the climatic signal recorded and maximise the resolving power of the data generated. In addition to records of the Cretaceous hot greenhouse interval, target sites will reveal the regional effects of the mid-Eocene–early Oligocene opening of the Tasman gateway and the Miocene–Pliocene restriction of the Indonesian gateway; to the present day, both gateways have important effects on global oceanography and climate.

The expedition also plans to drill deep holes into rifted basins that formed on the boundary between Australia, India and Antarctica before these continents drifted apart in the Cretaceous. The sedimentary record from these basins will provide information on the timing of the different

157

stages of the Gondwana breakup and complex magmatic history of the region at that time. Better knowledge of the magmatic history will allow exploration of possible temporal and causal relationships between high levels of volcanic activity during this time and the onset of Cretaceous greenhouse conditions in the region.

Regional setting of the Naturaliste Plateau and the Mentelle Basin

The Naturaliste Plateau (Figure 11.3) is a large (90,000 km^2) submarine plateau with water depths ranging mostly between 2,000 m and 3,000 m. From field analyses, dredging and previous DSDP drilling (Legs 26 and 28), the Naturaliste Plateau is believed to be underpinned by crystalline continental crust incised by Mesozoic and possibly even Palaeozoic rift basins and covered by an extensive carapace of volcanic rocks and related intrusives. The Mentelle Basin lies immediately to the east of the Naturaliste Plateau and is a large sedimentary basin potentially containing up to 12 km of sediments.

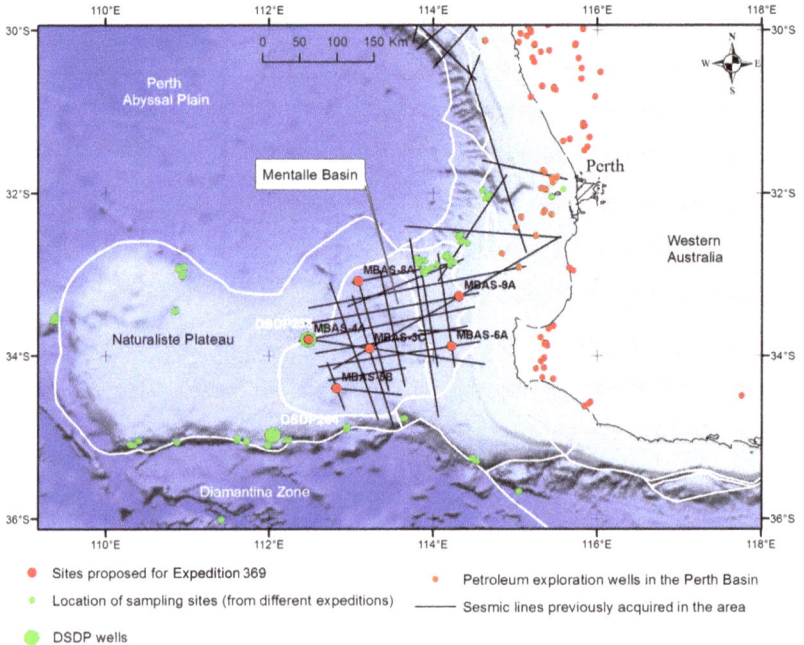

Figure 11.3. Location of the Naturaliste Plateau and the Mentelle Basin off the southwest Australian coast

Source: Geoscience Australia

The Naturaliste Plateau and the Mentelle Basin are key pieces in the Eastern Gondwana geodynamic reconstructions. The pre-breakup position of the Naturaliste Plateau against the Antarctic margin remains unclear, causing ongoing problems with Indian and Southern Ocean reconstructions.

The long history of continental extension between Greater India and Australia–Antarctica spanned the Palaeozoic and Mesozoic and created a set of depocentres along Australia's western margin, including the Mentelle Basin in the south. There were two separate rifting episodes along the western margin of Australia: Permian and Jurassic to Early Cretaceous, forming a complex system of interconnected rifted basins. Seismic interpretation suggests that remnants of the Permian rifts may be present beneath the eastern Mentelle Basin. During the second phase of rifting, in the Jurassic to Early Cretaceous, a series of deep depocentres formed on the Perth margin and up to 7 km of sediments were deposited in the western Mentelle Basin. At the very end of the Jurassic, breakup between Australia and India led to opening of the first Indian Ocean basin, the Argo Abyssal Plain, consisting of oceanic crust. By the Early Cretaceous (Valanginian), the breakup had advanced to the south and had reached the Naturaliste Plateau (Figure 11.4). This breakup at the southwestern corner of Australia was associated with significant intrusive and extrusive magmatism. Spectacular lava flows of the Bunbury Basalt and multiple volcanic facies recognised in seismic data on and above the Valanginian unconformity led to some researchers classifying this area as a volcanic margin. However, the origin and age(s) of volcanic sequences mapped on the Naturaliste Plateau and in the Mentelle Basin are not known; therefore, our understanding of the volcanic events in the area is still sketchy. In particular, potential links between volcanism on the southwestern corner of Australia and the Kerguelen Hotspot are debatable.

Rifting of Australia and Antarctica was protracted, taking approximately 50 million years before true sea-floor spreading was established in the mid-Palaeogene. Formation of this hyper-extended margin led to the unroofing of the crust and exposing of the mantle rocks in the Diamantina Zone. During this period, the western Mentelle Basin subsided quickly and accumulated a thick mid to upper Cretaceous succession. Drilling at DSDP Site 258 (located on the boundary between the Naturaliste Plateau and the Mentelle Basin, Figure 11.3) in 1972 sampled a 525 m section containing intervals of black shale believed to have been deposited during Cretaceous Oceanic Anoxic events. Seismic data acquired by Geoscience Australia indicates that the black shale section at Site 258 expands to the east into an almost 1 km thick sequence beneath the western Mentelle Basin. The new biostratigraphic interpretations of the excellently preserved foraminifera from site DSDP-258 (Figure 11.5) indicate that there is a high probability of a complete mid to late Cretaceous record, including both the OAE-1D and OAE-2 events.

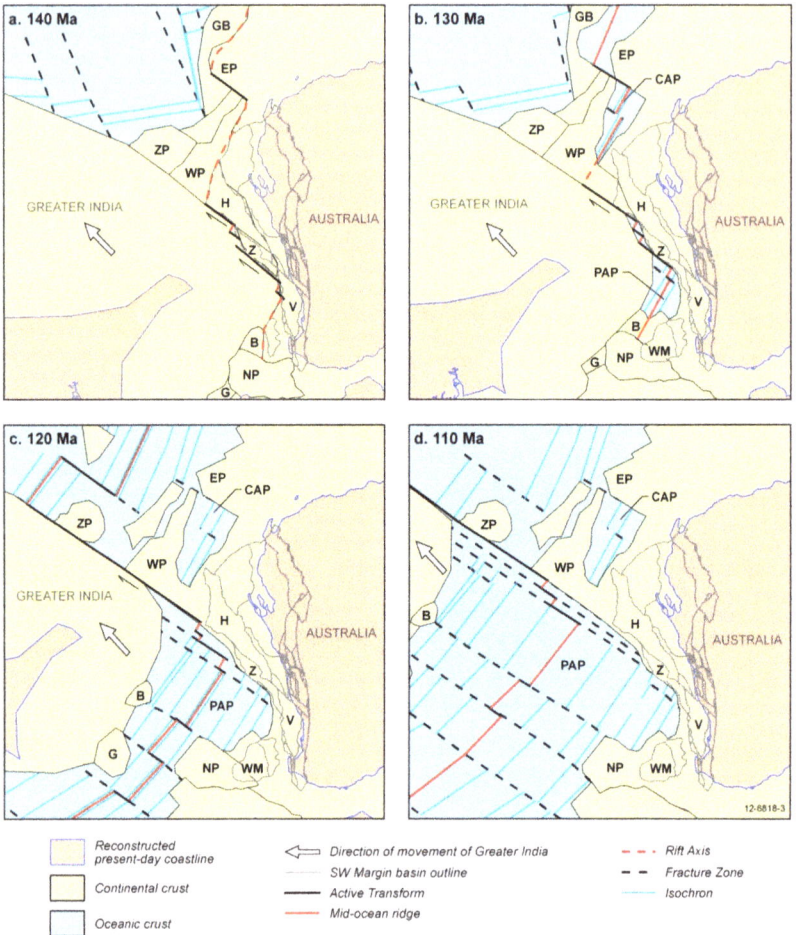

Figure 11.4. Regional plate reconstructions of the Australian southwest margin

Source: Modified from Gibbons et al. (2012) and Hall et al. (2013)

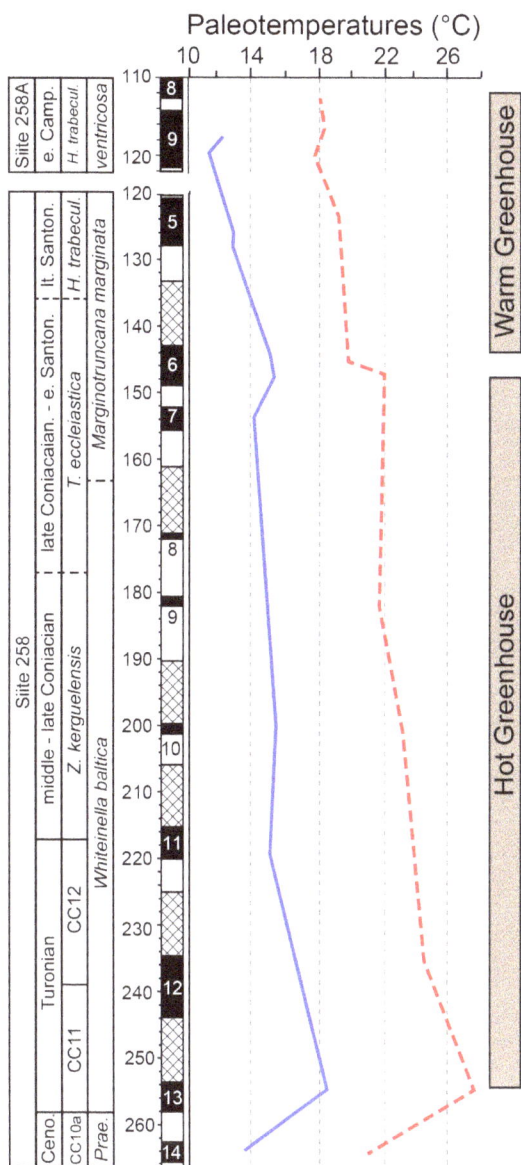

Figure 11.5. Stable isotope curves for DSDP Site 258 and Hole 258A based on data from Huber et al. (1995) and unpublished data (measured at the University of Missouri Stable Isotope Geochemistry Laboratory) relative to revised calcareous nannofossil and planktonic foraminiferal biostratigraphic zonal and age assignments

Cored intervals are labelled with black representing recovered core and white representing unrecovered core. The cross-hatched pattern represents intervals drilled without coring.

Source: Brian Huber

By the late Palaeogene, the Tasman gateway was open, allowing an alternative route for deep-water exchange between the Pacific and Indian Oceans, as by then the Indonesian gateway was closed by the northward movement of the Australian continent. Together with the opening of the Drake Passage, the Tasman gateway enabled the formation of the circumpolar current that dominates the Earth's climate to the present day. The location of the Naturaliste Plateau and the Mentelle Basin is ideal to sample the effects of changing deep-water ocean currents over this period.

Scientific rationale and drilling objectives

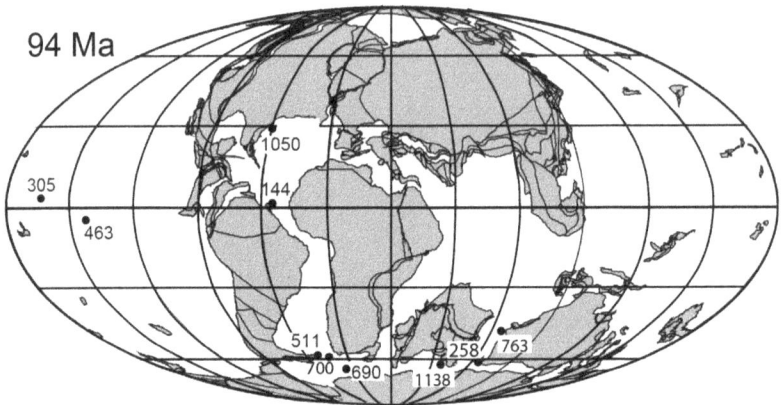

Figure 11.6. Late Cenomanian palaeogeographic map showing the location of DSDP Site 258 and select deep-sea sites yielding reliable oxygen isotope data

Source: Modified from Hay et al. (1999)

Constraint by observation of the mechanisms, feedbacks and time relationships that link climate dynamics from the polar regions to the tropics is of fundamental importance for reconstructing rapid climate change in the past. Understanding past climates is a critical test of understanding climates in general and, hence, to improving predictions for the future. High-resolution stratigraphic records from strategic locations around the globe, especially from the high latitude ocean, are essential to achieve this broader goal (Figure 11.6). Within this context, past periods of extreme warmth, such as the Cretaceous hot greenhouse (~100–89 Ma) and the Palaeocene-Eocene Thermal Maximum (55.8–55.6 Ma) have attracted increasing research interest over recent years. Results have provided often spectacular (and sometimes contradictory) insights into the mechanisms of natural short-term change in climate, biogeochemical cycling and ocean oxygenation. They have also demonstrated that short-term events are not independent from longer-term tectonic conditions (e.g. continental positions, depth of oceanic gateways and magmatic activity).

The Naturaliste Plateau and the Mentelle Basin provide a unique opportunity to study how the Earth's climate and oceans respond to elevated levels of atmospheric CO_2 and the role of local tectonic events in the onset of anoxic conditions. Key questions that will be addressed by scientific drilling here include:

- **Rise and collapse of the Cretaceous hot greenhouse.** What was the timing for onset of warming, and when were peak temperatures reached? When did cooling begin? What were the causes for these climatic transitions? How did the climate–ocean system and biota respond?

- **Oceanic Anoxic Events.** What were the relative roles of productivity and ocean circulation during these major carbon cycle perturbations that caused depletion of oxygen in vast areas of the world's oceans? Did any local factors (such as extensive volcanic activity or restricted circulation in semi-isolated basins) contribute to the onset of these conditions?

- **Cretaceous deep and intermediate water circulation.** Where were the main source regions for regional water masses in the southeast Indian Ocean, particularly during the peak of the hot greenhouse temperatures? How did these change during Gondwana breakup and climatic cooling?

- **Cenozoic palaeoceanography.** How did oceanographic conditions at the Naturaliste Plateau change during the opening of the Tasman and restriction of the Indonesian gateways?

- **Naturaliste Plateau and Mentelle Basin tectonic and depositional history.** What is the age of sediments in the syn-rift succession of the Mentelle Basin? What is the nature of basement beneath the Mentelle Basin? How many volcanic phases occurred and how did the environment change after the volcanism?

- **Gondwana breakup.** What was the position of the Naturaliste Plateau against the Antarctic terrains? When did the Naturaliste Plateau fully separate from India? When did the separation with Antarctica begin? How was the Diamantina Zone formed?

Project status

The planned drilling expedition to the Naturaliste Plateau and the Mentelle Basin, Expedition 369, is the result of a project proposal 760 submitted to IODP in 2014. After some revisions, the science proposal was accepted and recommended for drilling. In September 2015, the sites were approved by the Environment Protection and Safety Panel. The drilling leg was scheduled to start in September 2017. A site was added later in the Great Australian Bight, to drill the black shales of the Cenomanian–Turonian OAE 2 interval, which were dredged by Geoscience Australia scientists some years ago.

References

Gibbons, A.D., Barckhausen, U., van den Bogaard, P., Hoernle, K., Werner, R., Whittaker, J.M., and Muller, R.D., 2012. Constraining the Jurassic extent of Greater India: Tectonic evolution of the West Australian margin. *Geochemistry Geophysics Geosystems,* 13. doi.org/10.1029/2011GC003919

Hall, L.S., Gibbons, A.D., Bernardel, G., Whittaker, J.M., Nicholson, C., Rollet, N., and Mueller, R.D., 2013. Structural Architecture of Australia's Southwest Continental Margin and Implications for Early Cretaceous Basin Evolution, in Keep, M. and Moss, S.J. (eds), *The Sedimentary Basins of Western Australian IV. Proceedings of the Petroleum Exploration Society of Australia Symposium,* Perth, WA.

Hay, W.W., DeConto, R.M., Wold, C.N., Wilson, K.M., Voigt, S., Schulz, M., Wold-Rossby, A., Dullo, W.-C., Ronov, A.B., Balukhovsky, A.N., and Söding, E., 1999. Alternative global Cretaceous palaeogeography, in Barrera, E., and C. Johnson (eds), *Evolution of the Cretaceous Ocean-Climate System,* pp. 1–47, Geological Society of America Special Paper 332, Boulder, CO.

Huber, B.T., Hodell, D.A., and Hamilton, C.P., 1995. Middle–Late Cretaceous climate of the southern high latitudes: Stable isotopic evidence for minimal equator-to-pole thermal gradients. *Geological Society of America Bulletin,* 107: 1164–1191. doi.org/10.1130/0016-7606(1995)107<1164:MLCCOT>2.3. CO;2

Subduction Initiation and Palaeogene Climate (SIPC) in the Tasman Frontier, southwest Pacific: Forthcoming IODP Expedition 371

Rupert Sutherland, then GNS Science, now Victoria University of Wellington

(Co-proponents of IODP Proposal 832: G.R Dickens, Rice University; M. Gurnis, Caltech Seismological Labratory; J. Collot, Geological Survey of New Caledonia; M. Huber, Purdue University; **C.J. Hollis**, GNS Science; W. Stratford, GNS Science; S. Etienne, Geological Survey of New Caledonia; **B. Opdyke**, The Australian National University; H. Nishi, Tohuku University; E. Thomas, Yale University; **M. Seton**, University of Sydney; and W. Roest, IFREMER, France)

Subduction systems are the primary drivers of plate motions, mantle dynamics and global geochemical cycles, but little is known about how subduction starts. What are the initial conditions? How do forces and kinematics evolve? What are the short-term consequences and surface signatures: uplift, subsidence, deep-water sedimentary basins, convergence, extension and volcanism? Subduction initiation and changes in plate motion are linked because the largest driving and resisting tectonic forces occur within subduction zones.

Early Eocene onset of subduction in the western Pacific was accompanied by a profound global reorganisation of tectonic plates and a change in Pacific plate motion that is now manifest as the Emperor-Hawaii bend (Figure 11.7). The Izu-Bonin-Mariana and Tonga-Kermadec systems contain complementary information about this event, but the southwest Pacific has had little relevant drilling.

Our primary goal is to understand Tonga-Kermadec subduction initiation through recovery of Palaeogene sediment records from the Tasman Frontier region (Figure 11.8). Compilation of seismic-reflection data has led to new interpretations and hypotheses; hence drilling targets were identified, and four additional marine surveys were undertaken during the period 2013–15. High-quality sedimentary records with excellent fossil preservation are preserved in the region, due to optimal water depths of 1,000–4,000 m.

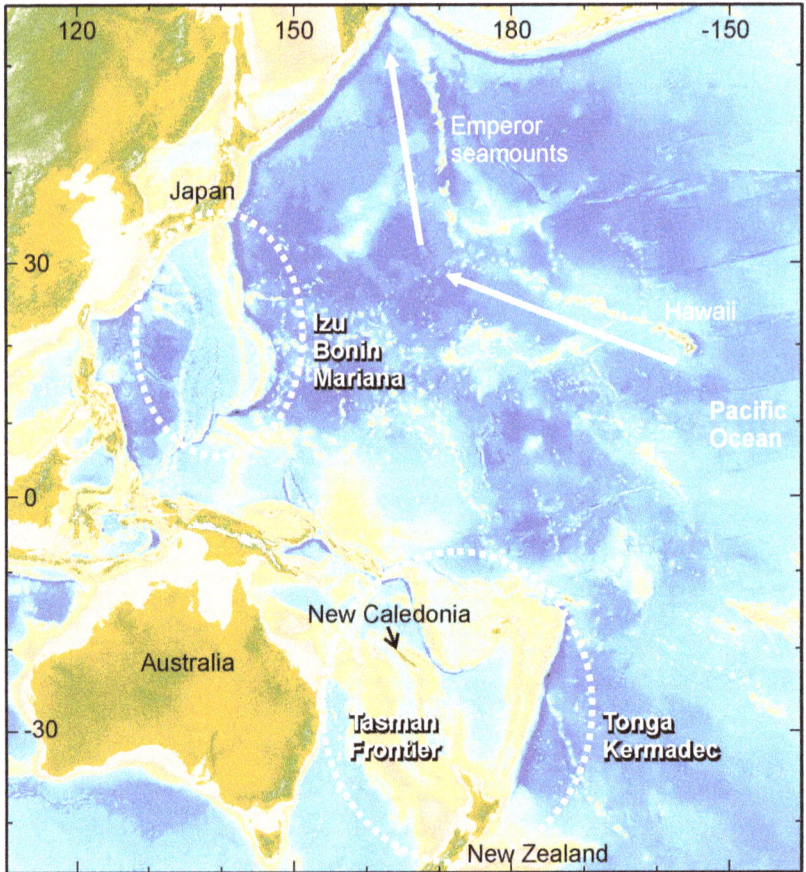

Figure 11.7. Location of Tonga-Kermadec subduction in the Tasman Frontier, southwest Pacific

The Tasman Frontier spans 3,000,000 km² between Australia, New Zealand and New Caledonia, and has bathymetric rises and troughs hosting sediment records of Tonga-Kermadec subduction initiation.

Source: Sutherland, R., Dickens, G.R., and Blum, P., 2016. *Expedition 371 Scientific Prospectus: Tasman Frontier Subduction Initiation and Palaeogene Climate.* International Ocean Discovery Program. dx.doi.org/10.14379/iodp.sp.371.2016

Eocene tectonic changes occurred at an important turning point in Cenozoic climate at ~50 Ma. Global warming through the Palaeocene–Eocene transition culminated in the Early Eocene Climate Optimum (~52–50 Ma), which was followed by an extended cooling trend that continued through the remaining Cenozoic. Understanding the nature and causes of this turning point will address an outstanding climate question: why does the Earth oscillate between multimillion-year greenhouse and icehouse climate states?

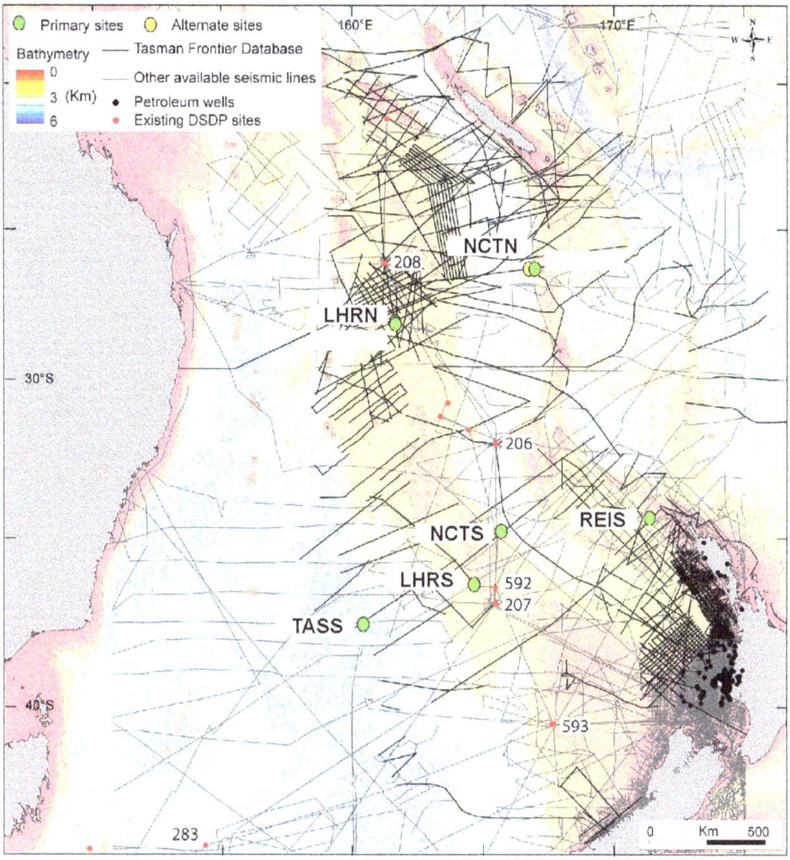

Figure 11.8. Available seismic data, and existing DSDP (red) and proposed drill sites (green)

Source: Sutherland, R., Dickens, G.R., and Blum, P., 2016. *Expedition 371 Scientific Prospectus: Tasman Frontier Subduction Initiation and Palaeogene Climate.* International Ocean Discovery Program. dx.doi.org/10.14379/iodp.sp.371.2016

Under very high pCO_2 conditions or high climate sensitivity, global climate models can accurately simulate Early Eocene warming in most regions, but not the extreme warmth reported in the southwest Pacific and Southern Oceans. Do these southwest Pacific proxy records point to undiscovered climate phenomena that amplify high-latitude temperatures during periods of global warmth? Would more accurate Tasman palaeobathymetry estimates alter models of ocean circulation and predict southern expansion of tropical currents?

Drilling at the proposed sites (Figure 11.8) will enable precise dating of deformed strata and the environmental changes associated with Tonga-Kermadec subduction initiation. Was subduction initiation synchronous throughout the western Pacific? How was the change in Pacific plate motion related to changes at the plate boundaries? Did compression occur synchronously in the near and far fields, relative to the proto-trench? What was the spatial and temporal pattern of vertical motions? Key observations may support the conjecture that a period of high-amplitude long-wavelength compression led to subduction initiation.

Our Palaeogene sediment studies will address these questions that are related to our primary science goal of understanding subduction initiation, and be significant in a key region for our secondary science goal of understanding Cenozoic global climate. The proposal may also have significant implications for how species dispersed through the south Pacific during Cenozoic time.

The SIPC proposal was strongly supported at the IODP Science Evaluation Panel meeting of January 2016, and approved by the *JOIDES Resolution* Facility Board in May 2016. The expedition is scheduled for mid-2018.

Unlocking the secrets of slow slip by drilling at the northern Hikurangi subduction margin, New Zealand: Including forthcoming IODP Expeditions 372 and 375

Laura Wallace, University of Texas at Austin and GNS Science, Lower Hutt

(On behalf of Demian Saffer, Pennsylvania State University, Pennsylvania; **Stuart Henrys**, GNS Science, Lower Hutt; **Phil Barnes**, NIWA, Wellington; and other members of the Hikurangi Margin working group)

Over the last decade, the discovery of episodic slow slip events (SSEs) at subduction margins around the globe has led to an explosion of new theories about fault rheology and slip behaviour along subduction megathrusts. The northern Hikurangi subduction margin offshore New Zealand is the only place on Earth where well-documented SSEs occur on a subduction interface within range of scientific drilling capabilities (Figures 11.9 and 11.10). Drilling, downhole measurements and sampling of the northern Hikurangi SSE source area will provide a unique opportunity to definitively test hypotheses for the physical conditions and rock properties leading to SSE occurrence and, ultimately, to unlock the secrets of slow slip. Northern Hikurangi subduction margin SSEs recur every one to two years, and thus provide an excellent setting to monitor changes in deformation rate and associated chemical and physical properties surrounding the SSE source area throughout a slow slip cycle.

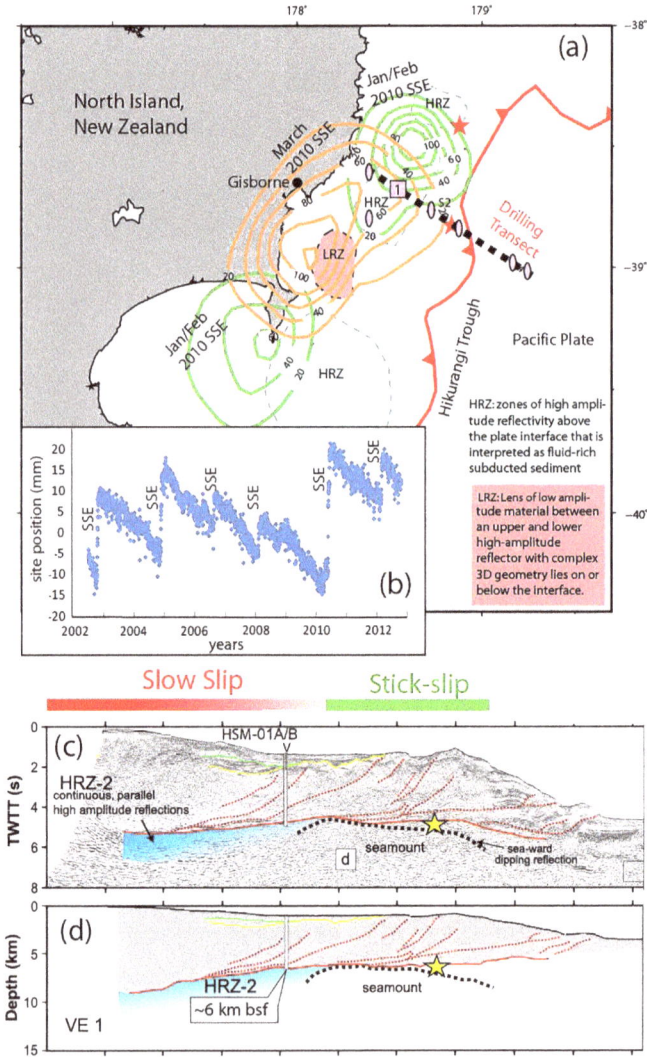

Figure 11.9. Tectonic setting and location of slip on the interface in the January/February (green contours) and the March/April (orange contours) 2010 SSEs and the reflective properties of the subduction interface at northern Hikurangi

Black dashed line with pink ellipses shows the location of the proposed riserless drilling transect (proposal 781A-Full), while the pink square shows the proposed riser drill site, HSM01-B (proposal 781B-Full). Inset Figure (b) shows the east component of the position time series for a continuous GPS site near Gisborne to demonstrate the repeatability of SSEs since they were first observed in 2002.

Source: Saffer, D.M., Wallace, L.M., and Petronotis, K., 2017. *Expedition 375 Scientific Prospectus: Hikurangi Subduction Margin Coring and Observatories*. International Ocean Discovery Program. doi.org/10.14379/iodp.sp.375.2017

To achieve this goal, a series of riserless and riser drilling proposals (781-CDP, 781A-Full, 781B-Full) targeting slow slip at the Hikurangi margin has been submitted to IODP over the last four years. This suite of proposals has involved numerous proponents from New Zealand, the US, Japan, Europe and Canada, and has included a large number of ANZIC (New Zealand) scientists as part of the lead proponent group (see table below). Proposal 781-CDP outlines the overall plan to use both riserless (*JOIDES Resolution*) and riser (*Chikyu*) drilling to discern the mechanisms behind subduction zone SSEs by drilling at northern Hikurangi. The riserless proposal (781A-Full) is focused on sampling the upper plate and subducting section via shallow drilling (400–1,500 m) and installation of borehole observatories (Figure 11.10). The primary aims of the riser phase (781B-Full) are to sample, log and conduct downhole measurements in the hanging wall and across the portion of the plate interface where SSEs occur at 5–6 km below the sea floor (Figure 11.10). Both 781A-Full and 781B-Full have received an 'excellent' ranking from the IODP Science Evaluation Panel and were forwarded to their respective facilities boards (*JOIDES Resolution* Facilities Board (JRFB) and *Chikyu* IODP Board (CIB)) for consideration for ranking and scheduling. Proposal 781A has given rise to Expeditions 372 and 375, the first of which will begin in late 2017.

Proposal title and number	Proponents
781-CDP: Multiphase Drilling Project: Unlocking the secrets of slow slip at the northern Hikurangi subduction margin.	L. Wallace (US/ANZIC), S. Henrys (ANZIC), P. Barnes (ANZIC), D. Saffer (USA), H. Tobin (USA), N. Bangs (USA), R. Bell (ECORD) and the Hikurangi margin working group (24 people).
781A-Full: Unlocking the secrets of slow slip by drilling at the northern Hikurangi subduction margin, NZ: Riserless drilling to sample the forearc and subducting plate.	D. Saffer (USA), P. Barnes (ANZIC), L. Wallace (US/ANZIC), S. Henrys (ANZIC), M. Underwood (USA), M. Torres (USA) and the Hikurangi margin working group (26 people).
781B-Full: Unlocking the secrets of slow slip at the northern Hikurangi subduction margin: Riser drilling to intersect the plate interface.	L. Wallace (US/ANZIC), Y. Ito (Japan), S. Henrys (ANZIC), P. Barnes (ANZIC), D. Saffer (USA), S. Kodaira (Japan), H. Tobin (USA), M. Underwood (USA), N. Bangs (USA), A. Fagereng (ECORD), H. Savage (USA), S. Ellis (ANZIC) and the Hikurangi margin working group (22 people).

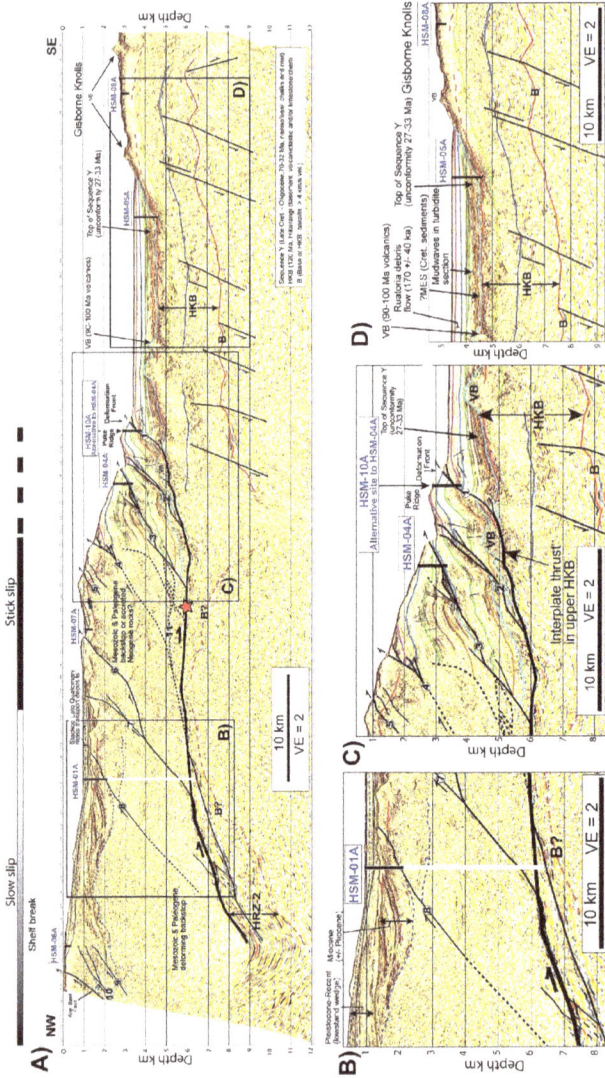

Figure 11.10. Interpretation of depth converted seismic profile 05CM-04 across the upper plate and subducting Pacific Plate east of Gisborne

The profile is co-located with the proposed/planned drilling transect. The bold black fault is the subduction interface. The proposed deep riser hole (co-located with HSM-01A) is denoted by the white heavy line, while the dark vertical lines show proposed riserless sites. Our latest trimmed-down plan prioritises *JOIDES Resolution* riserless drilling at HSM-05A, HSM-10A and HSM-01A. CORK subsurface hydrogeological observatories (funded by NSF) will be installed at HSM-01A and an alternate site for HSM-10A during riserless drilling in 2018.

Source: Saffer, D.M., Wallace, L.M., and Petronotis, K., 2017. *Expedition 375 Scientific Prospectus: Hikurangi Subduction Margin Coring and Observatories*. International Ocean Discovery Program. doi.org/10.14379/iodp.sp.375.2017

The riserless proposal (781A-Full) constitutes the first phase of the project and has advanced rapidly through the IODP system since its submission in late 2011. The JRFB has committed to scheduling 781A-Full for drilling as Expeditions 372 and 375 in the United States Fiscal Year 2018. Primary sites and a suite of alternate sites for riserless drilling have already been approved by IODP's Environmental Protection and Safety Panel (EPSP). The United States National Science Foundation (NSF) is funding the borehole observatory component of the project, which will be undertaken as a 'community experiment' with the data from the observatories being openly available to the public. The 3–4 planned riserless boreholes are designed to address three fundamental scientific objectives: (1) characterise the state and composition of the incoming plate and shallow plate boundary fault near the trench, which comprise the protolith and initial conditions for fault zone rock at greater depth; (2) characterise material properties, thermal regime and stress conditions in the upper plate above the SSE source region; and (3) install borehole observatory instruments to monitor a transect of holes above the SSE source, to measure temporal variations in deformation, fluid flow and seismicity.

The riser drilling proposal (781B-Full) has been named as one of several flagship *Chikyu* proposals and is awaiting consideration for drilling by the CIB. The riser borehole is designed to address two fundamental scientific objectives: (1) characterise the composition, mechanical properties and structural characteristics of the megathrust in the slow slip source area; and (2) characterise hydrological properties, thermal regime, stress and pore pressure state above and within the SSE source region. Together, these data will test a suite of hypotheses about the fundamental mechanics and behaviour of slow slip events, and their relationship to great subduction earthquakes. Without direct sampling of rocks from the SSE source and *in situ* measurements of physical properties (as proposed in 781B-Full), geoscientists are limited to speculation regarding the mechanisms that lead to SSEs. A group of American, New Zealand, Japanese and British scientists are in the process of pursuing funding to undertake a 3D seismic survey of the proposed riser drilling site, which is required in order to undertake riser drilling. A recent sea-floor geodetic and ocean bottom seismometer deployment suggests that large slow slip may actually extend to much shallower depths than originally thought, which suggests the possibility for an even shallower riser target (~2–3 km below the sea floor) to intersect a region of large slow slip on the Hikurangi subduction thrust. This result makes the possibility of riser drilling to intersect Hikurangi SSEs much more tractable than was previously thought.

Gateway to the Sub-Arc Mantle: Volatile flux, metal transport and conditions for early life: Forthcoming IODP Expedition 376

Cornel de Ronde, GNS Science, Lower Hutt

(On behalf of Wolfgang Bach, University of Bremen; **Richard Arculus**, The Australian National University; Susan Humphris, Woods Hole Oceanographic Institution; Ken Takai, JAMSTEC; Anna-Louise Reysenbach, Portland State University)

Hydrothermal systems hosted in submarine arc volcanoes differ substantially from those in spreading environments in that they commonly contain a large component of magmatic fluid. Our primary scientific goal is to discover the fundamental, underlying processes that drive these differences. This magmatic hydrothermal signature, coupled with the shallow depths of these volcanoes and high volatile contents, strongly influences the chemistry of the fluids and the resulting mineralisation, and likely has important consequences for the biota associated with these systems. Given the high metal contents and very acidic fluids, these hydrothermal systems are also thought to be important analogs of many porphyry copper and epithermal gold deposits mined on land.

Brothers Volcano on the Kermadec Arc (Figure 11.11) has been identified as the top candidate for arc volcano drilling to provide the missing link (i.e. the 3rd dimension) in our understanding of mineral deposit formation along arcs, the subseafloor architecture of these volcanoes and their related permeability, and the relationship between the discharge of magmatic fluids and the deep biosphere. The proposed drilling is highly likely to occur in 2018 and has four objectives:

i. characterising the sub-volcano, magma chamber-derived volatile phase to test model-based predictions that this is either a single-phase gas, or two-phase brine-vapour

ii. exploring the subseafloor distribution of base and precious metals and metalloids, and the reactions that have taken place along pathways to the sea floor

iii. quantifying the mechanisms and extent of fluid-rock interaction, the consequences for mass transfer of metals and metalloids into the ocean, and the role of magmatically derived carbon and sulphur species in mediating those fluxes

iv. assessing the diversity, extent and metabolic pathways of microbial life in an extreme, metal-toxic and acidic volcanic environment.

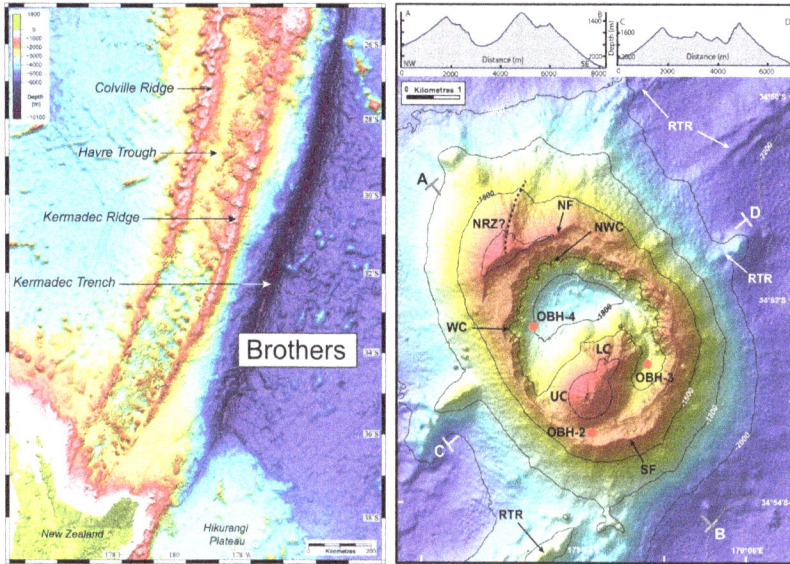

Figure 11.11. Bathymetry of the Kermadec Arc and Trench, and of Brothers Volcano

A. Bathymetry of the Kermadec Arc and Trench with major tectonic elements labelled. Brothers Volcano is located on the volcanic front in the southern half of the arc (from de Ronde et al., 2012). B. Detailed bathymetry of Brothers volcano and surrounds (modified after Embley et al., 2012). Dashed lines are structural ridges. Letters designate North fault (NF), South fault (SF), North rift zone (NRZ), Upper Cone (UC), and Lower Cone (LC), NW Caldera (NWC), W Caldera (WC) and regional tectonic ridge (RTR). Letters A–B and C–D are endpoints for the bathymetric cross sections shown in the top panels. The topographic cross-section 'A–B' is coincident with the seismic section 'Bro-03. Red dots mark the locations of the ocean bottom hydrophones (OBHs) referred to in the text. Contour interval is 200 m.

Source: GNS Science

We proposed a *JOIDES Resolution* expedition (IODP Proposal 818) to drill and log a series of sites at Brothers Volcano that will provide access to critical zones dominated by magma degassing and high-temperature hydrothermal circulation, over depth ranges regarded as crucial not only in the development of multiphase mineralising systems but also in identifying subsurface microbially habitable environments. We have identified and prioritised seven potential drill sites based on topographic slope, magnetic delineation of 'upflow' zones, and the location of hydrothermal vents that target all four hydrothermal fields that range from gas to seawater-dominated systems.

Our proposed drilling at Brothers Volcano (Figures 11.12, 11.13) is multidisciplinary and relates directly to challenges identified in the IODP New Science Plan 2013–2023, including two under the Earth Connections theme: one under the Earth in Motion theme and two under the Biosphere Frontiers Theme.

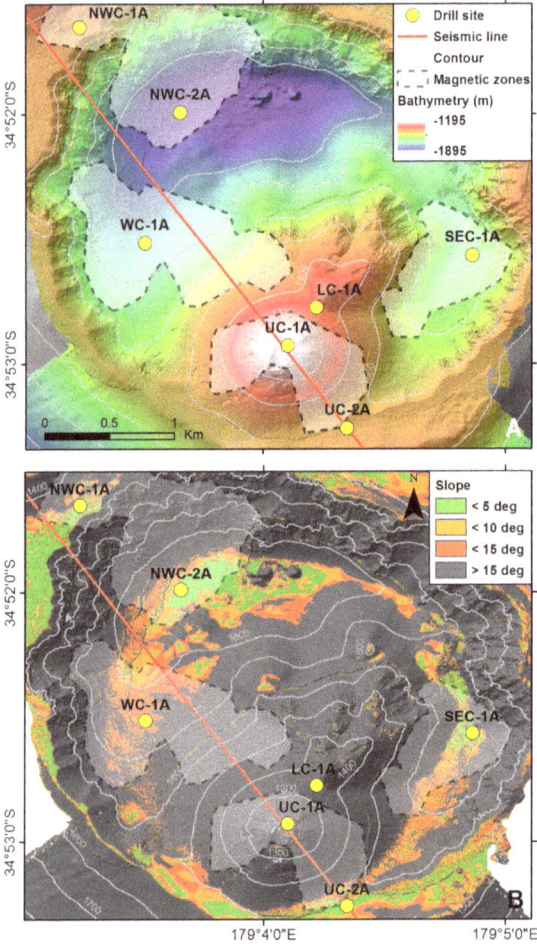

Figure 11.12. Proposed drill sites at Brothers Volcano

A. Location of proposed drill sites at Brothers Volcano. Transparent areas mark magnetic 'lows' inferred to be upflow zones within the Brothers hydrothermal system (Caratori Tontini et al., 2012). B. Slope map of Brothers. Drill sites NWC-1A, -2A and SEC-1A are sited on slopes of <5°; WC-1A on <10°; UC-1A and LC-1A <15° within a minimum circle area of 15 m up to 150 m. NWC-1A, WC-1A and UC-1A are primary sites with the others alternate sites. NWC, NW Caldera, UC, Upper Cone; LC, Lower Cone; SEC, SE Caldera, WC, West Caldera. Red line is seismic section 'Bro-03' shown in Fig. 11.13.

Source: GNS Science

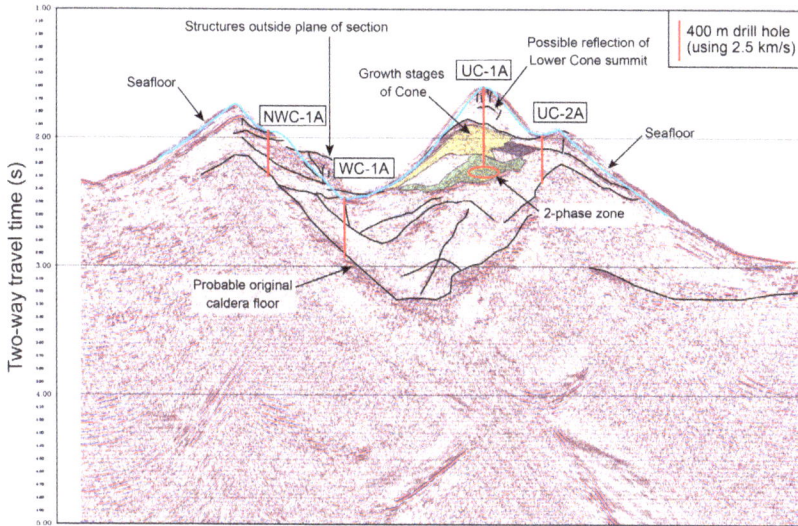

Figure 11.13. Seismic section along line 'Bro-3' for the three main drill holes (NWC-1A, WC-1A and UC-1A) and one backup hole (UC-2A)

Legend shows the length of a ~400 m hole based on a seismic velocity of 2.5 km/s; higher velocities will mean 'shallower' and lower velocities 'deeper' holes on the same section. Blue line is the sea floor derived from the bathymetric data. Hole WC-1A is slightly out of the plane of the section. Two-phase zone derived from OBHs.

Source: GNS Science and NIWA

The need for drilling: Why Brothers Volcano is the best primary site

Brothers Volcano has been the focus of multiple research voyages using surface ships, manned submersibles and remotely operated and autonomous underwater vehicles. Multidisciplinary research has focused on the volcanology, igneous petrology, geophysics, vent fluid geochemistry, numerical model simulation of fluid flow, mineral deposit formation and biological studies, culminating in the recent publication of an issue of *Economic Geology* largely devoted to this volcano (de Ronde et al., 2012).

The chance to drill separate but linked vent sites at a single location thus presents a uniquely important opportunity.

Brothers represents a window into the complicated hydrothermal systems that are found at submarine arc volcanoes, with a range of geological and structural settings, vent fluid chemistry, animals and microbes unheralded at any other site on the sea floor.

References

Caratori Tontini, F., Davy, B., de Ronde, C.E.J., Embley, R.W., Leybourne, M.I., and Tivey, M.A., 2012. Crustal magnetization of Brothers volcano, New Zealand, measured by autonomous underwater vehicles: Geophysical expression of a submarine hydrothermal system. *Economic Geology*, 107: 1571–1581. doi.org/10.2113/econgeo.107.8.1571

de Ronde, C.E.J., Butterfield, D.A., and Leybourne, M.I., 2012. Metallogenesis and Mineralization of Intraoceanic Arcs I: Kermadec arc—Introduction. *Economic Geology*, 107: 1521–1525. doi.org/10.2113/econgeo.107.8.1521

Embley, P.W., de Ronde, C.E.J., Merle, S.G., and Caratori Tontini, F., 2012. Detailed Morphology and Structure of an Active Submarine Arc Caldera: Brothers Volcano, Kermadec Arc, *Economic Geology*, 107: 1557–1570. doi.org/10.2113/econgeo.107.8.1557

Ocean–ice sheet interactions and West Antarctic Ice Sheet vulnerability (IODP Proposal 751-full): Forthcoming IODP Expedition 374

Rob McKay, Victoria University of Wellington

(On behalf of L. De Santis, CGS, Trieste; P. Bart, Louisiana State University; A. Shevenell, University of South Florida; T. Williams, Texas A&M University; **R. Levy**, GNS Science; L. Bartek, University of North Carolina; C. Sjunneskog, University of Otago; A. Orsi, Texas A&M University; S. Warny, Louisiana State University; R. DeConto, University of Amhurst; D. Pollard, Stanford University; Y. Suganuma, National Institute of Polar Research, Japan; and J. Hong, Korea Polar Research Institute)

The Southern Ocean is warming significantly, while Southern Hemisphere westerly winds have migrated southward and strengthened due to increasing atmospheric CO_2 concentrations and/or ozone depletion. These changes have been linked to thinning of Antarctic ice shelves and marine terminating glaciers. Numerical modelling combined with the results of geological drilling on Antarctica's continental margins indicates a fundamental role for oceanic heat in controlling marine ice sheet variability over at least the past 20 million years, particularly due to oceanic processes operating at the continental shelf edge. While ice sheet variability has been observed previously, sedimentary sequences from the outer continental shelf are still required to evaluate the exact oceanic processes that may have led to past marine ice sheet collapse events.

The Ross Sea drilling proposal (751-full, now Expedition 374; Figure 11.14) aims to drill a latitudinal and depth transect of six sites from the outer continental shelf and rise in the Eastern Ross Sea to resolve the relationship between climatic/oceanic change and West Antarctic Ice Sheet (WAIS) evolution through the past 23 million years. This location was selected because numerical ice sheet models indicate that it is highly sensitive to changes in ocean currents and heat, and thus the project is designed for optimal data–model integration to enable an improved understanding of the sensitivity of Antarctic Ice Sheet mass balance during warmer-than-present climates of the early Pliocene (~5.3–3.6 Ma) and middle Miocene (~18–14 Ma).

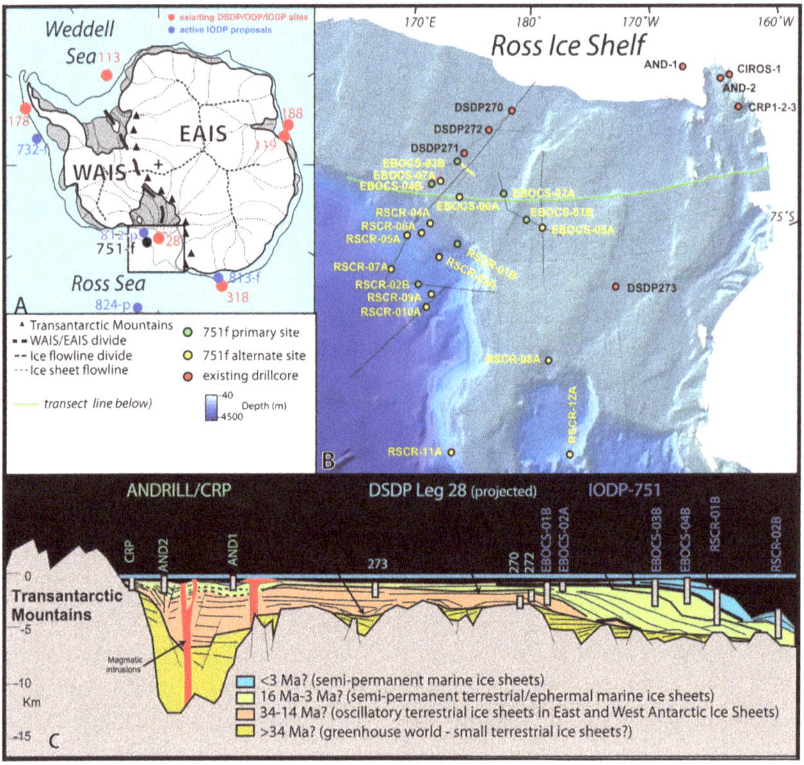

Figure 11.14. Ross Sea maps and seismic section

A. Antarctica with basic glaciology and previous/proposed DSDP/ODP/IODP. B. Ross Sea bathymetry with locations of proposed IODP 751-full sites (including alternates) and existing seismic network. C. Schematic of Ross Sea seismic stratigraphy and previous drilling highlighting continental to rise transect and linkages to previous drilling and evolution of WAIS.

Source: McKay, R.M., De Santis, L., and Kulhanek, D.K., 2017. Ross Sea West Antarctic Ice Sheet History: Ocean-ice sheet interactions and West Antarctic Ice Sheet vulnerability: Clues from the Neogene and Quaternary record of the outer Ross Sea continental margin. *IODP Scientific Prospectus*, 374. doi.org/10.14379/iodp.sp.374.2017

This latitudinal and depth transect from the Ross Sea continental shelf to rise is designed to achieve five scientific objectives:

1. Evaluate the contribution of West Antarctica ice volume variations to past sea-level change.

2. Reconstruct atmospheric and oceanic temperatures to identify past polar amplification and assess its forcings and feedbacks.

3. Assess the role of oceanic forcing (e.g. sea level and temperature) on WAIS stability/instability.

4. Understand natural ice age cycles by identify the sensitivity of WAIS to Earth's orbital configuration under a variety of climate boundary conditions.

5. Reconstruct Eastern Ross Sea bathymetry to examine relationships between sea-floor geometry and marine ice sheet instability.

The proposal is particularly relevant to the Australasian region as it continues New Zealand's long involvement in drilling on the inner Ross Sea continental shelf (e.g. Cape Robert Projects, ANDRILL projects) and deep drilling sites/targets collected offshore of Eastern New Zealand by the Ocean Drilling Program in 1998, which contain a record of the largest inflow of Antarctic Bottom Water (sourced from the Ross Sea) into the global ocean. The proposal was designed to integrate these older records into an unprecedented ice proximal to far-field oceanographic view of Antarctic cryosphere evolution in the southwest Pacific, and to determine how changes in this region have global implications for ocean heat transport. It also builds on recent Antarctic records collected south of Australia as part of the Wilkes Land expedition (IODP Leg 318), another of the three key regions of Antarctic Bottom Water (alongside the Ross and Weddell Seas). The proposal has been approved by the Science Evaluation Panel and scheduled by the JRFB for early 2018.

12
Broad costs and benefits of Australia's participation in IODP

From a scientific point of view, Australia's involvement in IODP empowers and enables our scientists to build a fundamental understanding of long-term issues in areas such as climate and oceanographic change, extremophile microbiology, sea-level rise, tsunami hazards and petroleum and mineral deposits. It also helps us understand our world's planetary dynamics, and better understand and manage the biodiversity and potential resources of our marine jurisdiction.

As a minor partner, we have little influence on the global IODP science framework – but we do have considerable influence in attracting expeditions to our part of the world. The Southern Hemisphere is a key region in studies of past global climate and oceanography, and there are very good reasons to bring other types of expeditions to this region. Such expeditions address global problems, but they also bring a dedicated team of 30 scientists from around the world to study our geoscience, not just for the two months of an expedition but for years to come.

There had been 17 DSDP/ODP expeditions and four IODP expeditions in our broad region by the end of 2013, with a present-day cost of perhaps US$200 million. This is the sort of sum that Australia and New Zealand could never find themselves, and our financial input to these expeditions was almost negligible in these terms.

Our community has long believed that these benefits of our involvement greatly exceed the costs. However, in order to have an independent view of the costs and benefits of Australia's involvement in IODP, the ANZIC Governing Council commissioned the Allen Consulting Group to provide a 94-page 'Review of Australia's participation in the Integrated Ocean Drilling Program'. This was finalised in March 2013 and is available from iodp.org.au/publications/independent-review-of-australian-participation-in-integrated-ocean-drilling-program/. Some findings are summaries below.

Section 6.1. Key findings of the review

Each successive phase of scientific ocean drilling (DSDP, ODP and IODP) has achieved significant scientific and technical results. This has been confirmed by reviews undertaken in the US, UK and Europe and by this study. Planning is already well underway for the next phase, the International Ocean Discovery Program, and a new Science Plan is agreed and revised operational arrangements are being put in place.

In assessing the benefits and costs of IODP, it is important to recognise that there are significant lags between the initiation of a proposal, actual drilling taking place, the analysis of results and publication of scientific papers. This means that during the period of IODP, benefits will arise that are attributable (in part or in whole) to previous phases and similarly the benefit of work done during IODP will flow through to the successive phase.

It is also important to recognise that, as with all science, a range of benefits from IODP would flow to Australia, usually with some lags, even if Australia was not a participating member. However, it is clear from this study that the nature, scale and timing of benefits to Australia would be quantifiably and qualitatively less.

Within the constraints of the data and information available, and recognising the issues of attribution and additionality associated with membership, it is clear that Australia's direct participation in IODP has generated a range of collaboration, capability building, scientific outputs, economic impacts and broader national interest benefits.

In considering the costs of Australia's membership, it is important to account for both direct (e.g. membership fees) and indirect (e.g. support provided by host institutions) costs. Although not practical to quantify, some costs incurred in one phase of scientific ocean drilling (e.g. the work done during IODP on the 2013–2013 Science Plan) may well benefit a successor phase. In total, it is estimated that the direct and indirect costs of Australia's participation in IODP from 2008–2013 totals approximately AU$14.8 million. Of this amount AU$3 million is in the form of in-kind contributions from researchers' host institutions that can also be regarded as leverage on the ARC grants for IODP membership, over and above the member institutions' ANZIC consortium fees.

Section 6.2. Overall assessment of benefits and costs

It is not practicable to quantify in dollar terms the net additional benefit to Australia of IODP membership because there are no robust and generally accepted indicators and/or methodology to do so. Attempting to calculate a standard cost–benefit ratio is therefore not appropriate. In considering the value of the benefits to Australia it is important to recognise that:

- the nature and impact of the benefits are very diverse, ranging from those of a public good nature (e.g. increased knowledge) to those that are commercial (e.g. exploration industry use); many of the benefits are longer term and cumulative in nature (e.g. networks, collaborations and enhanced human capital); the direct (e.g. port visits) and indirect (e.g. industry use) benefits are quite significant in their own right; and

- the potential benefits (e.g. a better understanding of natural hazards, discovery of hydrocarbon resources) are huge'

In summary, the assessment is whether the benefits of membership sufficiently outweigh the costs to justify the investment in IODP and the successor International Ocean Discovery Program. The overall conclusion of the review is that quantifiable and unquantifiable benefits to Australia of direct membership of the IODP consortium far exceed the modest costs of participation. Moreover, it would be detrimental to Australia's interests not to be a member of IODP and such participation is well aligned with current government policy as articulated in the 2012 National Science Investment Plan, the aspirations of the Australia in the Asian Century White Paper and Australia's policy of fostering international scientific collaborations.

The review has also concluded that, based on the value to Australia of participation in IODP and the expected benefits and costs of membership of the International Ocean Discovery Program, there is a strong case for continuing direct Australian membership.

This review and everything else in this volume make a compelling case that Australia is getting great value from its membership of the wonderful international scientific ocean drilling campaign that has been going for more than 45 years, by far the longest international geoscience program ever run.

A brief summary in 2016 would be that the Australian and New Zealand annual membership is 1 per cent of IODP's annual budget, but our scientific return is disproportionately high. The IODP has drilling assets worth US$1.1 billion and repositories holding more than 400 km of cores. All information goes into the public domain and a moratorium on recovered core material and related data is less than two years. Australian annual membership has direct economic return from each visit of IODP research vessels to Australian ports averaging US$1 million and indirect economic returns through petroleum exploration using IODP drilling results on our continental margins.

The renewal of ARC funding through until the end of 2020 at AU$2 million per annum is a very strong endorsement of IODP's value to Australia.

13

Major publications by ANZIC science party participants arising from 2008 to 2013 expeditions

First compiled by Maxine Kerr of The Australian National University in December 2015, and added to thereafter; it is current to early 2017. The listing is in numerical order of IODP expedtions, which did not always coincide with the timing of the expeditions.

We had no involvement in Expeditions 326, 327, 328 and 332, which had engineering aims, and in microbiological Expedition 336. The Baltic Sea alternative platform Expedition 347 occurred in late 2013, but our involvement started with the sampling and descriptive activities early in 2014; accordingly, it is not included here.

The names of ANZIC participants are shown in bold type. Note that IODP Preliminary Reports and Proceedings do not have page numbers.

Expedition 316: Chris Fergusson

Fergusson C.L., 2011. Data report: Clast counts and petrography of gravels from Site C0007, IODP Expedition 316, Nankai Trough. In Kinoshita, M., Tobin, H., Ashi, J., Kimura G., Lallemant, S., Screaton, E.J., Curewitz, D., Masago, H., Moe, K.T., and the IODP Expedition 314/315/316 Scientists. *Proceedings of the IODP*, 314/315/316: Washington, DC (Integrated Ocean Drilling Program Management International, Inc.), 20 pp.

Kimura, G., Screaton, E.J., Curewitz, D., Chester, F., Fabbri, O., **Fergusson, C.**, and the IODP Expedition 316 Scientists, 2008. NanTroSEIZE Stage 1A: NanTroSEIZE shallow megasplay and frontal thrusts. *IODP Preliminary Report*, 316.

Kinoshita, M., Tobin, H., Ashi, J., Kimura, G., Lallemant, S., Screaton, E.J., Curewitz, D., Masago, H., Moe, K.T., Chester, F., Fabbri, O., **Fergusson, C.**, and the IODP Expedition 314/315/316 Scientists, 2009. *Proceedings of the IODP*, 314/315/316: Washington, DC (Integrated Ocean Drilling Program Management International, Inc.).

Screaton, E., Kimura, G., Curewitz, D., Moore, G., Chester, F., Fabbri, O., **Fergusson, C.**, and the IODP Expedition 316 Scientific Party, 2009. Interactions between deformation and fluids in the frontal thrust region of the NanTroSEIZE transect offshore the Kii Peninsula, Japan: Results from IODP Expedition 316 Sites C0006 and C0007. *G3: Geochemistry, Geophysics, Geosystems*, 10. doi.org/10.1029/2009gc002713

Expedition 317: Bob Carter, Simon George, Greg Browne, Martin Crundwell, Kirsty Tinto

Crundwell, M.P., 2014. Pliocene to Late Eocene foraminiferal and bolboformid biostratigraphy of IODP Hole 317-U1352C, Canterbury Basin, New Zealand. *GNS Science Report 2014/15*, March 2014. 49 pp.

Dinarès-Turell, J., and **Tinto, K.**, 2014. Data report: paleomagnetism and rock magnetism of sediments from offshore Canterbury Basin, IODP Expedition 317. In Fulthorpe, C.S. Hoyanagi, K., Blum, P., and the Expedition 317 Scientists. *Proceedings of the IODP*, 317: Tokyo (Integrated Ocean Drilling Program Management International, Inc.).

Fulthorpe, C.S., Hoyanagi, K., Blum, P., Guèrin, G., Slagle, A.L., Blair, S.A., **Browne, G.H.**, **Carter, R.M.**, Ciobanu, M.-C., Claypool, G.E., **Crundwell, M.P.**, Dinarès-Turell, J., Ding, X., **George, S.**, Hepp, D.A., Jaeger, J., Kawagata, S., Kemp, D.B., Kim, Y.-G., Kominz, M.A., Lever, H., Lipp, J.S., Marsaglia, K.M., McHugh, C.M., Murakoshi, N., Ohi, T., Pea, L., Richaud, M., Suto, I., Tanabe, S., **Tinto, K.J.**, Uramoto, G., and Yoshimura, T., 2010. Canterbury Basin Sea Level: Global and local controls on continental margin stratigraphy. *IODP Preliminary Report*, 317.

Fulthorpe, C.S., Hoyanagi, K., Blum, P., Guèrin, G., Slagle, A.L., Blair, S.A., **Browne, G.H.**, **Carter, R.M.**, Ciobanu, M.-C., Claypool, G.E., **Crundwell, M.P.**, Dinarès-Turell, J., Ding, X., **George, S.**, Hepp, D.A., Jaeger, J., Kawagata, S., Kemp, D.B., Kim, Y.-G., Kominz, M.A., Lever, H., Lipp, J.S., Marsaglia, K.M., McHugh, C.M., Murakoshi, N., Ohi, T., Pea, L., Richaud, M., Suto, I., Tanabe, S., **Tinto, K.J.**, Uramoto, G., and Yoshimura, T., 2011. IODP Expedition 317: Exploring the Record of Sea-Level Change off New Zealand. *Scientific Drilling*, 12: 4–14. doi.org/10.5194/sd-12-4-2011

Fulthorpe, C.S., Hoyanagi, K., Blum, P., Guèrin, G., Slagle, A.L., Blair, S.A., **Browne, G.H.**, **Carter, R.M.**, Ciobanu, M.-C., Claypool, G.E., **Crundwell, M.P.**, Dinarès-Turell, J., Ding, X., **George, S**., Hepp, D.A., Jaeger, J., Kawagata, S., Kemp, D.B., Kim, Y.-G., Kominz, M.A., Lever, H., Lipp, J.S., Marsaglia, K.M., McHugh, C.M., Murakoshi, N., Ohi, T., Pea, L., Richaud, M., Suto, I., Tanabe, S., **Tinto, K.J.**, Uramoto, G., and Yoshimura, T., 2011. *Proceedings of the IODP*, 317: Tokyo (Integrated Ocean Drilling Program Management International, Inc.).

Land M., Wust R.A.J., Robert C., and **Carter R.M.**, 2010. Plio-Pleistocene paleoclimate in the Southwest Pacific as reflected in clay mineralogy and particle size at ODP Site 1119, SE New Zealand. *Marine Geology*, 274: 165–176. doi.org/10.1016/j.margeo.2010.04.001

Villaseñor T., Jaeger J.M., Marsaglia K.M., and **Browne G.H.**, 2015. Evaluation of the relative roles of global versus local sedimentary controls on Middle to Late Pleistocene formation of continental margin strata, Canterbury Basin, New Zealand. *Sedimentology*, 62, 1118–1148. doi.org/10.1111/sed.12181

Expedition 318: Kevin Welsh, Robert McKay

Bijl, P.K., Bendle, J.A.P., Bohtay, S.M., Pross, J., Schouten, S., Tauxe, L., Stickley, C.E., **McKay, R.M.**, Rohl, U., Olney, M., Sluijs, A., Escutia, C., Brinkhius, H., **Welsh, K.**, and the IODP Expedition 318 Scientists, 2013. Eocene cooling linked to early flow across the Tasmanian Gateway. *Proceedings of the National Academy of Sciences of the United States of America*, 110(24): 9645–9650. doi.org/10.1073/pnas.1220872110

Brinkhuis, H., Escutia, C., Klaus, A., Fehr, A., Williams, T., Bendle, J.A.P., Bijl, P.K., Bohaty, S.M., Carr, S.A., Dunbar, R.B., Gonzàlez, J.J., Hayden, T.G., Iwai, M., Jimenez-Espejo, F.J., Katsuki, K., Kong, G.S., **McKay, R.M.**, Nakai, M., Olney, M.P., Passchier, S., Pekar, S.F., Pross, J., Riesselman, C., Röhl, U., Sakai, T., Shrivastava, P.K., Stickley, C.E., Sugisaki, S., Tauxe, L., Tuo, S., van de Flierdt, T., **Welsh, K.**, and Yamane, M., 2010. Wilkes Land Glacial History: Cenozoic East Antarctic Ice Sheet evolution from Wilkes Land margin sediments. *IODP Preliminary Report*, 318: 1–101.

Cook C.P., Van De Flierdt T., Williams T., Hemming S.R., Iwai M., Kobayashi M., Jimenez-Espejo F.J., Escutia C., Gonzalez J.J., Khim B.-K., **McKay R.M.**, Passchier S., Bohaty S.M., Riesselman C.R., Tauxe L., Sugisaki S., Galindo A.L., Patterson M.O., Sangiorgi F., Pierce E.L., Brinkhuis H., Klaus A., Fehr A., Bendle J.A.P., Bijl P.K., Carr S.A., Dunbar R.B., Flores J.A., Hayden T.G., Katsuki K., Kong G.S., Nakai M., Olney M.P., Pekar S. ., Pross J., Rohl U., Sakai T., Shrivastava P.K., Stickley C.E., Tuo S., **Welsh K.**, and Yamane M., 2013. Dynamic behaviour of the East Antarctic ice sheet during Pliocene warmth. *Nature Geoscience*, 6: 765–769. doi.org/10.1038/ngeo1889

Escutia, C., Brinkhuis, H., Klaus, A., Fehr, A., Williams, T., Bendle, J.A.P., Bijl, P.K., Bohaty, S.M., Carr, S.A., Dunbar, R.B., Gonzàlez, J.J., Hayden, T.G., Iwai, M., Jimenez-Espejo, F.J., Katsuki, K., Kong, G.S., **McKay, R.M.**, Nakai, M., Olney, M.P., Passchier, S., Pekar, S.F., Pross, J., Riesselman, C., Röhl, U., Sakai, T., Shrivastava, P.K., Stickley, C.E., Sugisaki, S., Tauxe, L., Tuo, S., van de Flierdt, T., **Welsh, K.**, and Yamane, M., 2011. IODP Expedition 318: from Greenhouse to Icehouse at the Wilkes Land Antarctic margin. *Scientific Drilling*, 12: 15–23. doi.org/10.5194/sd-12-15-2011

Escutia, C., Brinkhuis, H., Klaus, A., Fehr, A., Williams, T., Bendle, J.A.P., Bijl, P.K., Bohaty, S.M., Carr, S.A., Dunbar, R.B., Gonzàlez, J.J., Hayden, T.G., Iwai, M., Jimenez-Espejo, F.J., Katsuki, K., Kong, G.S., **McKay, R.M.**, Nakai, M., Olney, M.P., Passchier, S., Pekar, S.F., Pross, J., Riesselman, C., Röhl, U., Sakai, T., Shrivastava, P.K., Stickley, C.E., Sugisaki, S., Tauxe, L., Tuo, S., van de Flierdt, T., **Welsh, K.**, and Yamane, M., 2011. *Proceedings of the IODP*, 318: Tokyo (Integrated Ocean Drilling Program Management International, Inc.).

Houben, A.J.P., Bijl, P.K., Pross, J., Bohaty, S.M., Passchier, S., Stickley, C.E., Röhl, U., Sugisaki, S., Tauxe, L., van de Flierdt, T., Olney, M., Sangiorgi, F., Sluijs, A., Escutia, C., Brinkhuis, H., Klaus, A., Fehr, A., Williams, T., Bendle, J.A.P., Carr, S.A., Dunbar, R.B., Gonzàlez, J.J., Hayden, T.G., Iwai, M., Jimenez-Espejo, F.J., Katsuki, K., Kong, G.S., **McKay, R.M.**, Nakai, M., Pekar, S.F., Pross, J., Riesselman, C., Sakai, T., Shrivastava, P.K., Tuo, S., **Welsh, K.**, and Yamane, M., 2013. Reorganization of Southern Ocean plankton ecosystem at the onset of Antarctic glaciation. *Science*, 340(6130): 341–344. doi.org/10.1126/science.1223646

Mackintosh A.N., Verleyen E., O'Brien P.E., White D.A., Jones R.S., **McKay R.**, Dunbar R., Gore D.B., Fink D., Post A.L., Miura H., Leventer A., Goodwin I., Hodgson D.A., Lilly K., Crosta X., Golledge N.R., Wagner B., Berg S., Van Ommen T., Zwartz D., Roberts S.J., Vyverman W., and Masse G., 2014. Retreat history of the East Antarctic Ice Sheet since the Last Glacial Maximum. *Quaternary Science Reviews*, 100: 10–30. doi.org/10.1016/j.quascirev.2013.07.024

Orejola, N., Passchier, S., Brinkhuis, H., Escutia, C., Klaus, A., Fehr, A., Williams, T., Bendle, J.A.P., Bijl, P.K., Bohaty, S.M., Carr, S.A., Dunbar, R.B., Gonzàlez, J.J., Hayden, T.G., Iwai, M., Jimenez-Espejo, F.J., Katsuki, K., Kong, G.S., **McKay, R.M.**, Nakai, M., Olney, M.P., Pekar, S.F., Pross, J., Riesselman, C., Röhl, U., Sakai, T., Shrivastava, P.K., Stickley, C.E., Sugisaki, S., Tauxe, L., Tuo, S., van de Flierdt, T., **Welsh, K.**, and Yamane, M., 2014. Sedimentology of lower Pliocene to upper Pleistocene diamictons from IODP Site U1358, Wilkes Land margin, and implications for East Antarctic Ice Sheet dynamics. *Antarctic Science*, 26(2): 183–192. doi.org/10.1017/S0954102013000527

Pant, N.C., Biswas, P., Shrivastava, P.K., Bhattacharya, S., Verma, K., Pandey, M., Brinkhuis, H., Escutia, C., Klaus, A., Fehr, A., Williams, T., Bendle, J.A.P., Bijl, P.K., Bohaty, S.M., Carr, S.A., Dunbar, R.B., Gonzàlez, J.J., Hayden, T.G., Iwai, M., Jimenez-Espejo, F.J., Katsuki, K., Kong, G.S., **McKay, R.M.**, Nakai, M., Olney, M.P., Passchier, S., Pekar, S.F., Pross, J., Riesselman, C., Röhl, U., Sakai, T., Stickley, C.E., Sugisaki, S., Tauxe, L., Tuo, S., van de Flierdt, T., **Welsh, K.**, and Yamane, M., 2013. *Provenance of Pleistocene sediments from Site U1359 of the Wilkes Land IODP Leg 318 – evidence for multiple sourcing from the East Antarctic Craton and Ross Orogen*. Geologic Society, London, Special Publications, 381.

Patterson, M.O., **McKay, R.**, Naish, T., Escutia, C., Jimenez-Espejo, F.J., Raymo, M.E., Meyers, S.R., Tauxe, L., Brinkhuis, H., Klaus, A., Fehr, A., Williams, T., Bendle, J.A.P., Bijl, P.K., Bohaty, S.M., Carr, S.A., Dunbar, R.B., Gonzàlez, J.J., Hayden, T.G., Iwai, M., Katsuki, K., Kong, G.S., Nakai, M., Olney, M.P., Passchier, S., Pekar, S.F., Pross, J., Riesselman, C., Röhl, U., Sakai, T., Shrivastava, P.K., Stickley, C.E., Sugisaki, S., Tuo, S., van de Flierdt, T., **Welsh, K.**, and Yamane, M., 2014. Orbital forcing of the East Antarctic ice sheet during the Pliocene and Early Pleistocene. *Nature Geoscience*, 7(11): 841–847. doi.org/10.1038/ngeo2273

Pross, J., Contreras, L., Bijl, P.K., Greenwood, D.R., Bohaty, S.M., Schouten, S., Bendle, J.A., Rohl, U., Tauxe, L., Raine, J.I., Huck, C.E., van de Flierdt, T., Jamieson, S.S.R., Stickley, C.E., van de Schootbrugge, B., Escutia, C., Brinkhuis, H., **Welsh, K.**, **McKay, R.**, and the IODP Expedition 318 Scientists, 2012. Persistent near-tropical warmth on the Antarctic continent during the early Eocene epoch. *Nature*, 488: 73–77. doi.org/10.1038/nature11300

Stocchi P., Escutia C., Houben A.J.P., Vermeersen B.L.A., Bijl P.K., Brinkhuis H., Deconto R.M., Galeotti S., Passchier S., Pollard D., Brinkhuis H., Escutia C., Klaus A., Fehr A., Williams T., Bendle J.A.P., Bijl P.K., Bohaty S.M., Carr S.A., Dunbar R.B., Flores J.A., Gonzàlez J.J., Hayden T.G., Iwai M., Jimenez-Espejo F.J., Katsuki K., Kong G.S., **McKay R.M.**, Nakai M., Olney M.P., Passchier S., Pekar S.F., Pross J., Riesselman C., Röhl U., Sakai T., Shrivastava P.K., Stickley C.E., Sugisaki S., Tauxe L., Tuo S., Van De Flierdt T., **Welsh K.**, and Yamane M., 2013. Relative sea-level rise around East Antarctica during Oligocene glaciation. *Nature Geoscience*, 6: 380–384. doi.org/10.1038/ngeo1783

Tauxe, L., Stickley, C.E., Sugisaki, S., Bijl, P.K., Bohaty, S.M., Brinkhuis, H., Escutia, C., Flores, J.A., Houben, A.J.P., Iwai, M., Jimenez-Espejo, F., **McKay, R.**, Passchier, S., Pross, J., Riesselman, C.R., Rohl, U., Sangiorgi, F., **Welsh, K.**, Klaus, A., Fehr, A., Bendle, J.A.P., Dunbar, R., Gonzalez, J., Hayden, T., Katsuki, K., Olney, M.P., Pekar, S.F., Shrivastava, P.K., van de Flierdt, T., Williams, T., and Yamane, M., 2012. Chronostratigraphic framework for the IODP Expedition 318 cores from the Wilkes Lan Margin: constraints for paleoceanographic reconstruction. *Paleoceanography*, 27(2): 2214–2214. doi.org/10.1029/2012PA002308

Verma, K., Bhattacharya, S., Biswas, P., Shrivastava, P.K., Pandey, M., Pant, N.C., Brinkhuis, H., Escutia, C., Klaus, A., Fehr, A., Williams, T., Bendle, J.A.P., Bijl, P.K., Bohaty, S.M., Carr, S.A., Dunbar, R.B., Gonzàlez, J.J., Hayden, T.G., Iwai, M., Jimenez-Espejo, F.J., Katsuki, K., Kong, G.S., **McKay, R.M.**, Nakai, M., Olney, M.P., Passchier, S., Pekar, S.F., Pross, J., Riesselman, C., Röhl, U., Sakai, T., Stickley, C.E., Sugisaki, S., Tauxe, L., Tuo, S., van de Flierdt, T., **Welsh, K.**, and Yamane, M., 2014. Clay mineralogy of the ocean sediments from the wilkes land margin, east antarctica: Implications on the paleoclimate, provenance and sediment dispersal pattern. *International Journal of Earth Sciences : Geologische Rundschau*, 103(8): 2315–2326. doi. org/10.1007/s00531-014-1043-4

Welsh, K., 2011. From Greenhouse to Icehouse on the Wilkes Land Margin. *Australasian Science*, 32(9): 15–17.

Expedition 319: Gary Huftile

Hayman, N.W., Byrne T.B., McNeill L.C., Kanagawa K., Kanamatsu T., Browne C.M., Schleicher A.M., and **Huftile G.J.**, 2012. Structural evolution of an inner accretionary wedge and forearc basin initiation, Nankai margin, Japan. *Earth and Planetary Science Letters*, 353–354: 163–172. doi.org/10.1016/j. epsl.2012.07.040

Lin W., Doan M.-L., Moore J.C., McNeill L., Byrne T.B., Ito T., Saffer D., Conin M., Kinoshita M., Sanada Y., Moe K.T., Araki E., Tobin H., Boutt D., Kano Y., Hayman N.W., Flemings P., **Huftile G.J.**, Cukur D., Buret C., Schleicher A.M., Efimenko N., Kawabata K., Buchs D.M., Jiang S., Kameo K., Horiguchi K., Wiersberg T., Kopf A., Kitada K., Eguchi N., Toczko S., Takahashi K., and Kido Y. 2010. Present-day principal horizontal stress orientations in the Kumano forearc basin of the southwest Japan subduction zone determined from IODP NanTroSEIZE drilling Site C0009. *Geophysical Research Letters*, 37. doi.org/10.1029/2010gl043158

McNeill, L., Saffer, D., Byrne, T., Araki, E., Toczko, S., Eguchi, N., Takahashi, K., Kyaw Thu, M., Sanada, Y., Boutt, D., Buchs, D., Buret, C., Conin, M., Cukur, D., Doan, M.-L., Efimenko, N., Flemings, P., Hayman, N., Horiguchi, K., **Huftile, G.**, and the IODP Expedition 319 Scientists, 2010. IODP Expedition 319, NanTroSEIZE Stage 2: first IODP riser drilling operations and observatory installation towards understanding subduction zone seismogenesis. *Scientific Drilling*, 10: 4–13. doi.org/10.5194/sd-10-4-2010

Moore, J.C., Chang, C., McNeill, L., Thu, M.K., Yamada, Y., and **Huftile, G.**, 2011. Growth of borehole breakouts with time after drilling: Implications for state of stress, NanTroSEIZE transect, SW Japan. *G3: Geochemistry, Geophysics, Geosystems*, 12(4). doi.org/10.1029/2010GC003417

Saffer, D., McNeill, L., Araki, E., Byrne, T., Eguchi, N., Toczko, S., Takahashi, K., Kyaw Thu, M., Sanada, Y., Boutt, D., Buchs, D., Buret, C., Conin, M., Cukur, D., Doan, M.-L., Efimenko, N., Flemings, P., Hayman, N., Horiguchi, K., **Huftile, G.**, and the IODP Expedition 319 Scientists, 2009. NanTroSEIZE Stage 2: NanTroSEIZE riser/riserless observatory. *IODP Preliminary Report*, 319.

Saffer, D., McNeill, L., Byrne, T., Araki, E., Toczko, S., Eguchi, N., Takahashi, K., Kyaw Thu, M., Sanada, Y., Boutt, D., Buchs, D., Buret, C., Conin, M., Cukur, D., Doan, M.-L., Efimenko, N., Flemings, P., Hayman, N., Horiguchi, K., **Huftile, G.**, and the IODP Expedition 319 Scientists, 2010. *Proceedings of the IODP*, 319: Tokyo (Integrated Ocean Drilling Program Management International, Inc.).

Expedition 320: Christian Ohneiser

Channell, J.E.T., **Ohneiser, C.**, Yamamoto, Y., and Kesler, M.S., 2013. Oligocene-Miocene magnetic stratigraphy carried by biogenic magnetite at sites U1334 and U1335 (equatorial Pacific Ocean). *G3: Geochemistry, Geophysics, Geosystems*, 14(2).

Guidry, E.P., Richter, C., Acton, G.D., Channell, J.E.T., Evans, H.F., **Ohneiser, C.**, Yamamoto, Y., and Yamazaki, T., 2012. Oligocene–Miocene magnetostratigraphy of deep-sea sediments from the equatorial Pacific (IODP Site U1333). *Geological Society, London, Special Publications*, 373, 13–27. doi.org/10.1144/SP373.7

Lyle, M., Pälike, H., Nishi, H., Raffi, I., Gamage, K., Klaus, A., Acton G., Anderson L., Backman J., Baldauf J., Beltran C., Bohaty S.M., Bownpaul, Busch W., Channell J.E.T., Chun C.O.J., Delaney M., Dewangan P., Dunkley Jones T., Edgar K.M., Evans H., Fitch P., Foster G.L., Gussone N., Hasegawa H., Hathorne E.C., Hayashi H., Herrle J.O., Holbourn A., Hovan S., Hyeong K., Iijima K., Ito T., Kamikuri S.-I., Kimoto K., Kuroda J., Leon-Rodriguez L., Malinverno A., Moore Jr T.C., Murphy B.H., Murphy D.P., Nakamura H., Ogane K., **Ohneiser C.**, and the IODP Expeditions 320/321 Science Party, 2010. The Pacific Equatorial Age Transect, IODP Expeditions 320 and 321: building a 50-million-year-long environmental record of the equatorial Pacific Ocean. *Scientific Drilling*, 9: 4–15. doi.org/10.5194/sd-9-4-2010

Ohneiser, C., Acton, G., Channell, J.E.T., Wilson, G.S., Yamamoto, Y., Yamazaki, T., 2013. A middle Miocene relative paleointensity record from the Equatorial Pacific. *Earth and Planetary Science Letters,* 374: 227–238. doi.org/10.1016/j.epsl.2013.04.038

Pälike H., Lyle M.W., Nishi H., Raffi I., Ridgwell A., Gamage K., Klaus A., Acton G., Anderson L., Backman J., Baldauf J., Beltran C., Bohaty S.M., Bownpaul, Busch W., Channell J.E.T., Chun C.O.J., Delaney M., Dewangan P., Dunkley Jones T., Edgar K.M., Evans H., Fitch P., Foster G.L., Gussone N., Hasegawa H., Hathorne E.C., Hayashi H., Herrle J.O., Holbourn A., Hovan S., Hyeong K., Iijima K., Ito T., Kamikuri S.-I., Kimoto K., Kuroda J., Leon-Rodriguez L., Malinverno A., Moore Jr T.C., Murphy B.H., Murphy D.P., Nakamura H., Ogane K., **Ohneiser C.,** Richter C., Robinson R., Rohling E.J., Romero O., Sawada K., Scher H., Schneider L., Sluijs A., Takata H., Tian J., Tsujimoto A., Wade B.S., Westerhold T., Wilkens R., Williams T., Wilson P.A., Yamamoto Y., Yamamoto S., Yamazaki T., and Zeebe R.E., 2012. A Cenozoic record of the equatorial Pacific carbonate compensation depth. *Nature,* 488: 609–614. doi.org/10.1038/nature11360

Pälike, H., Lyle, M., Nishi, H., Raffi, I., Gamage, K., Klaus, A., Evans, H., Williams, T., Acton, G.D., Bown, P., Delaney, M., Dunkley Jones, T., Edgar, K., Fitch, P., Gussone, N., Herrle, J., Hyeong, K., Kamikuri, S., Kuroda, J., Leon-Rodriguez, L., Moore, T., Murphy, B., Nakamura, H., **Ohneiser, C.,** and the IODP Expedition 320/321 Scientists, 2010. *Proceedings of the IODP,* 320/321: Tokyo (Integrated Ocean Drilling Program Management International, Inc.).

Pälike, H., Nishi, H., Lyle, M., Raffi, I., Klaus, A., Gamage, K., Evans, H., Williams, T., Acton, G.D., Bown, P., Delaney, M., Dunkley Jones, T., Edgar, K., Fitch, P., Gussone, N., Herrle, J., Hyeong, K., Kamikuri, S., Kuroda, J., Leon-Rodriguez, L., Moore, T., Murphy, B., Nakamura, H., **Ohneiser, C.,** and the IODP Expedition 320/321 Scientists, 2009. Pacific Equatorial Transect. *IODP Preliminary Report,* 320.

Westerhold, T., Röhl, U., Wilkens, R., Pälike, H., Lyle, M., Dunkley-Jones, T., Bown, P., Moore, T., Kamikuri, S., Acton, G., **Ohneiser, C.,** Yamamoto, Y., Richter, C., Fitch, P., Scher, H., Liebrand, D., and Expedition 320/321 Scientists, 2012. Revised composite depth scales and integration of IODP Sites U1331–U1334 and ODP Sites 1218–1220. *Proceedings of the IODP,* Volume 320/321. doi.org/10.2204/iodp.proc.320321.201.2012

Yamamoto Y., Yamazaki T., Acton G.D., Richter C., Guidry E.P., and **Ohneiser C.**, 2014. Palaeomagnetic study of IODP Sites U1331 and U1332 in the equatorial Pacific – extending relative geomagnetic palaeointensity observations through the Oligocene and into the Eocene. *Geophysical Journal International*, 196: 694–711. doi.org/10.1093/gji/ggt412

Expedition 322: John Moreau

Saito, S., Underwood, M.B., Kubo, Y., Sanada, Y., Chiyonobu, S., Destrigneville, C., Dugan, B., Govil, P., Hamada, Y., Heuer, V., Hüpers, A., Ikari, M., Kitimara, Y., Kutterolf, S., Labanieh, S., **Moreau, J.**, and the IODP Expedition 322 Scientists, 2010. *Proceedings of the IODP*, 322: Tokyo (Integrated Ocean Drilling Program Management International, Inc.).

Torres, M.E., Cox, T., Hong, W.-L., McManus, J., Sample, J.C., Destrigneville, C., Gan, H.M., Gan, H.Y. and **Moreau, J.W.**, 2015. Crustal fluid and ash alteration impacts on the biosphere of Shikoku Basin sediments, Nankai Trough, Japan. *Geobiology*, 13: 562–580. doi.org/10.1111/gbi.12146

Underwood, M.B., Saito, S., Kubo, Y., Sanada, Y., Chiyonobu, S., Destrigneville, C., Dugan, B., Govil, P., Hamada, Y., Heuer, V., Hüpers, A., Ikari, M., Kitimara, Y., Kutterolf, S., Labanieh, S., **Moreau, J.**, and the IODP Expedition 322 Scientists, 2009. NanTroSEIZE Stage 2: subduction inputs. *IODP Preliminary Report*, 322.

Underwood, M.B., Saito, S., Kubo, Y., Sanada, Y., Chiyonobu, S., Destrigneville, C., Dugan, B., Govil, P., Hamada, Y., Heuer, V., Hüpers, A., Ikari, M., Kitimara, Y., Kutterolf, S., Labanieh, S., **Moreau, J.**, and the IODP Expedition 322 Scientists, 2010. IODP Expedition 322 drills two sites to document inputs to the Nankai Trough Subduction Zone. *Scientific Drilling*, 10: 14–25. doi.org/10.5194/sd-10-14-2010

Expedition 323: Kelsie Dadd

Dadd, K.A., and Foley, K., 2016. A shape and compositional analysis of ice-rafted debris in core from IODP Leg 323 in the Bering Sea. *Deep Sea Research II*. doi.org/10.1016/j.dsr2.2016.02.007

Ravelo, C., Takahashi, K., Alvarez Zarikian, C., Guèrin, G., Liu, T., Aiello, I., Asahi, H., Bartoli, G., Caissie, B., Chen, M., Colmenaro-Hidalgo, E., Cook, M., **Dadd, K.**, and the IODP Expedition 323 Scientists, 2010. Bering Sea paleoceanography: Pliocene-Pleistocene paleoceanography and climate history of the Bering Sea. *IODP Preliminary Report*, 323.

Takahashi, K., Ravelo, A.C., Alvarez Zarikian, C.A., Guèrin, G., Liu, T., Aiello, I., Asahi, H., Bartoli, G., Caissie, B., Chen, M., Colmenaro-Hidalgo, E., Cook, M., **Dadd, K.**, and the IODP Expedition 323 Scientists, 2011. *Proceedings of the IODP*, 323: Tokyo (Integrated Ocean Drilling Program Management International, Inc.).

Takahashi, K., Ravelo, A.C., and Alvarez Zarikian, C., Guèrin, G., Liu, T., Aiello, I., Asahi, H., Bartoli, G., Caissie, B., Chen, M., Colmenaro-Hidalgo, E., Cook, M., **Dadd, K.**, and the IODP Expedition 323 Scientists, 2011. IODP Expedition 323 – Pliocene and Pleistocene paleoceanographic changes in the Bering Sea. *Scientific Drilling*, 11: 4–13. doi.org/10.5194/sd-11-4-2011

Wehrmann L.M., Risgaard-Petersen N., Schrum H.N., Walsh E.A., Huh Y., Ikehara M., Pierre C., D'hondt S., Ferdelman T.G., Ravelo A.C., Takahashi K., Zarikian C.A. , Aiello, I., Asahi, H., Bartoli, G., Caissie, B., Chen, M., Colmenero-Hidalgo, E., Cook, M., **Dadd, K.**, and the IODP Expedition 323 Scientific Party, 2011. Coupled organic and inorganic carbon cycling in the deep subseafloor sediment of the northeastern Bering Sea Slope (IODP Exp. 323). *Chemical Geology*, 284: 251–261. doi.org/10.1016/j.chemgeo.2011.03.002

Expedition 324: David Murphy

Heydolph, K., **Murphy, D.T.**, Geldmacher, J., Romanova, I.V., Greene, A., Hoernle, K., Weis, D., and Mahoney, J., 2014. Plume versus plate origin for the Shatsky Rise oceanic plateau (NW Pacific): Insights from Nd, Pb and Hf isotopes. *Lithos*, 200–201: 49–63. doi.org/10.1016/j.lithos.2014.03.031

Sager, W.W., Sano, T., Geldmacher, J., Iturrino, G, Evans, H., Almeev, R., Ando, A., Carvallo, C., Delacour, A., Greene, A.R., Harris, A.C., Herrmann, S., Heydolph, K., Hirano, N., Ishikawa, A., Kang, M.H., Koppers, A.A.T., Li, S., Littler, K., Mahoney, J., Matsubara, N., Miyoshi, M., **Murphy, D.T.**, and the IODP Expedition 324 Scientists, 2010. Testing plume and plate models of ocean plateau formation at Shatsky Rise, northwest Pacific Ocean. *IODP Preliminary Report*, 324.

Sager, W.W., Sano, T., Geldmacher, J., Iturrino, G, Evans, H., Almeev, R., Ando, A., Carvallo, C., Delacour, A., Greene, A.R., Harris, A.C., Herrmann, S., Heydolph, K., Hirano, N., Ishikawa, A., Kang, M.H., Koppers, A.A.T., Li, S., Littler, K., Mahoney, J., Matsubara, N., Miyoshi, M., **Murphy, D.T.**, and the IODP Expedition 324 Scientists, 2011. IODP Expedition 324: ocean drilling at Shatsky Rise gives clues about oceanic plateau formation. *Scientific Drilling*, 12: 24–31. doi.org/10.5194/sd-12-24-2011

Sager, W.W., Sano, T., Geldmacher, J., Iturrino, G, Evans, H., Almeev, R., Ando, A., Carvallo, C., Delacour, A., Greene, A.R., Harris, A.C., Herrmann, S., Heydolph, K., Hirano, N., Ishikawa, A., Kang, M.H., Koppers, A.A.T., Li, S., Littler, K., Mahoney, J., Matsubara, N., Miyoshi, M., **Murphy, D.T.**, and the IODP Expedition 324 Scientists, 2010. *Proceedings of the IODP*, 324: Tokyo (Integrated Ocean Drilling Program Management International, Inc.).

Expedition 325: Jody Webster, Michael Gagan, Tezer Esat

Abbey E., **Webster J.M.**, and Beaman R.J., 2011. Geomorphology of submerged reefs on the shelf edge of the Great Barrier Reef: The influence of oscillating Pleistocene sea-levels. *Marine Geology*, 288: 61–78. doi.org/10.1016/j.margeo.2011.08.006

Abbey, E., **Webster, J.M.**, Braga, J.C., Jacobsen, G.E., Thorogood, G., Thomas, A.L., Camoin, G., Reimer, P.J., and Potts, D.C., 2013. Deglacial mesophotic reef demise on the Great Barrier Reef. *Palaeogeography, Palaeoclimatology, Palaeoecology*, 392: 473–494. doi.org/10.1016/j.palaeo.2013.09.032

Bridge, T.C.L., Fabricius, K.E., Bongaerts, P., Wallace, C.C., Muir, P.R., Done, T.J., and **Webster, J.M.**, 2012. Diversity of Scleractinia and Octocorallia in the mesophotic zone of the Great Barrier Reef, Australia. *Coral Reefs*, 31(1): 179–189. doi.org/10.1007/s00338-011-0828-1

Camoin, G., and **Webster, J.**, 2014. Coral reefs and sea-level change, in Stein, R., Blackman, D., Inagaki, F., and Christian-Larsen, H. (eds), *Earth and Life Processes Discovered from Subseafloor Environment – A decade of Science Achieved by the Integrated Ocean Drilling Program (IODP)*. Elsevier: Amsterdam/New York. doi.org/10.1016/b978-0-444-62617-2.00015-3

Camoin G.F., and **Webster J.M.**, 2015. Coral reef response to Quaternary sea-level and environmental changes: State of the science. *Sedimentology*, 62: 401–428. doi.org/10.1111/sed.12184

Felis, T., McGregor, H.V., Linsley, B.K., Tudhope, A.W., Gagan, M.K., Suzuki, A., Inoue, M., Thomas, A.L., **Esat, T.M.**, Thompson, W.G., Tiwari, M., Potts, D.C., Mudelsee, M., Yokoyama, Y., and **Webster, J.M.**, 2014. Intensification of the meridional temperature gradient in the Great Barrier Reef following the Last Glacial Maximum. *Nature Communications*, 5: 4102. doi.org/10.1038/ncomms5102

Gischler, E., Thomas, A.L., Droxler, A.W., **Webster, J.M.**, Yokoyama, Y., and Schöne, B.R., 2013. Microfacies and diagenesis of older Pleistocene (pre-Last Glacial Maximum) reef deposits, Great Barrier Reef, Australia (IODP Expedition 325): A quantitative approach. *Sedimentology*, 60(6): 1432–1466. doi.org/10.1111/sed.12036

Harper, B.B., Puga-Bernabéu, Á., Droxler, A.W., **Webster, J.M.**, Gischler, E., Tiwari, M., Lado-Insua, T., Thomas, A.L., Morgan, S., Jovane, L., and Röhl, U., 2015. Mixed carbonate-siliciclastic sedimentation along the Great Barrier Reef upper slope: a challenge to the reciprocal sedimentation model. *Journal of Sedimentary Research*, 85(9): 1019–1036. doi.org/10.2110/jsr.2015.58.1

Hinestrosa G., **Webster J.M.**, Beaman R.J., and Anderson L.M., 2014. Seismic stratigraphy and development of the shelf-edge reefs of the Great Barrier Reef, Australia. *Marine Geology*, 353: 1–20. doi.org/10.1016/j.margeo.2014.03.016

Hinestrosa, G., **Webster, J.M.**, and Beaman, R.J., 2016. Postglacial sediment deposition along a mixed carbonate-siliciclastic margin: New constraints from the drowned shelf-edge reefs of the Great Barrier Reef, Australia. *Palaeogeography, Palaeoclimatology, Palaeoecology*, 446: 168–185. doi.org/10.1016/j.palaeo.2016.01.023

Insua, T.L., Hamel, L., Moran, K., Anderson, L.M., and **Webster, J.M.**, 2015. Advanced classification of carbonate sediments based on physical properties. *Sedimentology*, 62(2): 590–606. doi.org/10.1111/sed.12168

Puga-Bernabéu, A., **Webster, J.M.**, Beaman, R.J., Reimer, P.J., and Renema, W., 2014. Filling the gap: A 60 ky record of mixed carbonate-siliciclastic turbidite deposition from the Great Barrier Reef. *Marine and Petroleum Geology*, 50: 40–50. doi.org/10.1016/j.marpetgeo.2013.11.009

Renema, W., Beaman, R.J., **Webster, J.M.**, 2013. Mixing of relict and modern tests of larger benthic foraminifera on the Great Barrier Reef shelf margin. *Marine Micropaleontology*, 101: 68–75. doi.org/10.1016/j.marmicro.2013.03.002

Webster, J.M., Yokoyama, Y., Cotterill, C., Anderson, L., Green, S., Bourillot, R., Braga, J.C., Drowler, A., **Esat, T.**, Felis, T., Fujita, K., **Gagan, M.**, and the IODP Expedition 325 Scientists, 2010. Great Barrier Reef environmental changes: the last deglacial sea level rise in the South Pacific: offshore drilling northeast Australia. *IODP Preliminary Report*, 325. doi.org/10.2204/iodp.pr.325.2010

Webster, J.M., Yokoyama, Y., Cotterill, C., Anderson, L., Green, S., Bourillot, R., Braga, J.C., Drowler, A., **Esat, T.**, Felis, T., Fujita, K., **Gagan, M.**, and the Expedition 325 Scientists, 2011. Great Barrier Reef environmental changes. *Proceedings of the IODP*, 325: Tokyo (Integrated Ocean Drilling Program Management International, Inc.). doi.org/10.2204/iodp.proc. 325.2011

Woodroffe C.D., and **Webster J.M.**, 2014. Coral reefs and sea-level change. *Marine Geology*, 352: 248–267. doi.org/10.1016/j.margeo.2013.12.006

Yokohama, Y., **Webster, J.M.**, Cotterill, C., Braga, J.C., Jovane, L., Mills, H., Morgan, S., Suzuki, A., Anderson, L., Green, S., Bourillot, R., Drowler, A., **Esat, T.**, Felis, T., Fujita, K., **Gagan, M.**, and the IODP Expedition 325 Scientists, 2011. IODP Expedition 325: The Great Barrier Reef reveals past sea-level, climate, and environmental changes since the last Ice Age. *Scientific Drilling*, 12: 32–45. doi.org/10.2204/iodp.sd.12.04.2011

Expedition 329: Jill Lynch

D'Hondt, S., Inagaki, F., Alvarez Zarikian, C., Abrams., L.J., Dubois, N., Engelhardt, T., Evans, H., Ferdelman, T., Gribsholt, B., Harris, R.N., Hoppie, B.W., Hyun, J.-H., Kallmeyer, J., Kim, J., **Lynch, J.E.**, McKinley, C.C., Mitsunobu, S., Morono, Y., Murray, R.W., Pockalny, R., Sauvage, J., Shimono, T., Shiraishi, F., Smith, D.C., Smith-Duque, C.E., Spivack, A.J., Steinsbu, B.O., Suzuki, Y., Szpak, M., Toffin, L., Uramoto, G., Yamaguchi, Y.T., Zhang, G., Zhang, X.-H., and Ziebis, W., 2015. Presence of oxygen and aerobic communities from sea floor to basement in deep-sea sediments. *Nature Geoscience*, 8: 299–304. doi.org/10.1038/ngeo2387

D'Hondt, S., Inagaki, F., Alvarez Zarikian, C., Evans, H., Bubois, N., Engelhardt, T., Ferdelman, T., Gribsholt, B., Harris, R.N., Hoppie, B.W., Hyun, J.-H., Kallmeyer, J., Kim, J., **Lynch, J.E.**, and the IODP Expedition 329 Scientists, 2011. South Pacific Gyre subseafloor life. *IODP Preliminary Report*, 329.

D'Hondt, S., Inagaki, F., Alvarez Zarikian, C.A., Evans, H., Bubois, N., Engelhardt, T., Ferdelman, T., Gribsholt, B., Harris, R.N., Hoppie, B.W., Hyun, J.-H., Kallmeyer, J., Kim, J., **Lynch, J.E.**, and the IODP Expedition 329 Science Party, 2013. IODP Expedition 329: Life and habitability beneath the seafloor of the South Pacific Gyre. *Scientific Drilling*, 15: 4–10. doi.org/10.5194/sd-15-4-2013

D'Hondt, S., Inagaki, F., Alvarez Zarikian, C.A., Evans, H., Bubois, N., Engelhardt, T., Ferdelman, T., Gribsholt, B., Harris, R.N., Hoppie, B.W., Hyun, J.-H., Kallmeyer, J., Kim, J., **Lynch, J.E.**, and the Expedition 329 Scientists, 2011. *Proceedings of the IODP*, 329: Tokyo (Integrated Ocean Drilling Program Management International, Inc.).

Expedition 330: Benjamin Cohen, David Buchs

Koppers, A.A.P., Yamazaki, T., Geldmacher, J., Anderson, L., Beier, C., **Buchs, D.M.**, Chen, L.-H., **Cohen, B.E.**, and the IODP Expedition 330 Scientists, 2011. Louisville Seamount Trail: Implications for geodynamic mantle flow models and the geochemical evolution of primary hotspots. *IODP Preliminary Report*, 330.

Koppers, A.A.P., Yamazaki, T., Geldmacher, J., Anderson, L., Beier, C., **Buchs, D.M.**, Chen, L.-H., **Cohen, B.E.**, and the IODP Expedition 330 Scientific Party, 2013. IODP Expedition 330: Drilling the Louisville Seamount Trail in the SW Pacific. *Scientific Drilling*, 15: 11–22. doi.org/10.5194/sd-15-11-2013

Koppers, A.A.P., Yamazaki, T., Geldmacher, J., Anderson, L., Beier, C., **Buchs, D.M.**, Chen, L.-H., **Cohen, B.E.**, and the IODP Expedition 330 Scientists, 2012. *Proceedings of the IODP*, 330: Tokyo (Integrated Ocean Drilling Program Management International, Inc.).

Koppers, A.A.P., Yamazaki, T., Geldmacher, J., Gee, J.S., Pressling, N., Hoshi, H., Anderson, L., Beier, C., **Buchs, D.M.**, Chen, L.-H., **Cohen, B.E.**, Deschamps, F., Dorais, M.J., Ebuna, D., Ehmann, S., Fitton, J.G., Fulton, P.M., Ganbat, E., Hamelin, C., Hanyu, T., Kalnins, L., Kell, J., Machida, S., Mahoney, J.J., Moriya, K., Nichols, A.R.L., Rausch, S., Sano, S., Sylvan, J.B., and Williams, R., 2012. Limited latitudinal mantle plume motion for the Louisville hotspot. *Nature Geoscience*, 5: 911–917. doi.org/10.1038/ngeo1638

Nichols, A.R.L., Beier, C., Brandl, P.A., **Buchs, D.M.**, and Krumm, S.H., 2014. Geochemistry of volcanic glasses from the Louisville Seamount Trail (IODP Expedition 330): Implications for eruption environments and mantle melting. *G3: Geochemistry, Geophysics, Geosystems*, 15(5): 1718–1738. doi.org/10.1002/2013GC005086

Expedition 331: Chris Yeats

Ishibashi J-I., Miyoshi, Y., Inoue, H., **Yeats, C.**, Hollis, S.P., Corona, J.C., Bowden, S., Yang, S., Southam, G., Masaki, Y. and Hartnett, H., 2013. Subseafloor structure of a submarine hydrothermal system within volcaniclastic sediments: a modern analogue for 'Kuroko-type' VMS deposits, in Jonsson, E, et al. (eds) Mineral deposit research for a high-tech world. *Proceedings of the 12th Biennial SGA Meeting*, 12–15 August 2013, Uppsala, Sweden, pp. 542–544.

Mottl, M., Takai, K., Nielsen, S.H.H., Birrien, J.-L., Bowden, S., Brandt, L., Breuker, A., Corona, J.C., Eckert, S., Hartnett, H., Hollis, S.P., House, C.H., Ijiri, A., Ishibashi, J., Masaki, Y., McAllister, S., McManus, J., Moyer, C., Nishizawa, M., Noguchi, T., Nunoura, T., Southam, G., Yanagawa, K., Yang, S., and **Yeats, C.**, 2010. Deep hot biosphere. *IODP Preliminary Report*, 331.

Takai, K., Mottl, M.J., Nielsen, S.H.H., Birrien, J.-L., Bowden, S., Brandt, L., Breuker, A., Corona, J.C., Eckert, S., Hartnett, H., Hollis, S.P., House, C.H., Ijiri, A., Ishibashi, J., Masaki, Y., McAllister, S., McManus, J., Moyer, C., Nishizawa, M., Noguchi, T., Nunoura, T., Southam, G., Yanagawa, K., Yang, S., and **Yeats, C.**, 2012. IODP Expedition 331: strong and expansive subseafloor hydrothermal activities in the Okinawa Trough. *Scientific Drilling*, 13: 19–27. doi.org/10.5194/sd-13-19-2012

Takai, K., Mottl, M.J., Nielsen, S.H., Birrien, J.-L., Bowden, S., Brandt, L., Breuker, A., Corona, J.C., Eckert, S., Hartnett, H., Hollis, S.P., House, C.H., Ijiri, A., Ishibashi, J., Masaki, Y., McAllister, S., McManus, J., Moyer, C., Nishizawa, M., Noguchi, T., Nunoura, T., Southam, G., Yanagawa, K., Yang, S., and **Yeats, C.**, 2011. *Proceedings of the IODP*, 331: Tokyo (Integrated Ocean Drilling Program Management International, Inc.).

Yeats, C.J., and Hollis, S.P., 2013. Actively forming Kuroko-style massive sulfide mineralisation and hydrothermal alteration at Iheya North, Okinawa Trough. Goldschmidt 2013, Florence, Italy, 25–30 August. *Mineralogical Magazine*, 77(5). doi.org/10.1180/minmag.2013.077.5.25

Yeats, C., Hollis, S., LaFlamme, C., Halfpenny, A., Fiorentini, M., Corona, J.-C., Southam, G., Herrington, R., and Spratt, J., 2016. Actively forming Kuroko-type VMS mineralization at Iheya North, Okinawa Trough, Japan: new geochemical, petrographic and δ34S isotope results. Mineral Deposit Studies Group, Winter Meeting 2016. University College Dublin, 4–7 January 2016. *Applied Earth Science* (2016). doi.org/10.1080/03717453.2016.1166680

Yeats, C.J., Hollis, S.P., Halfpenny, A., Corona, J.-C., LaFlamme, C., Southam, G., Fiorentini, M., Herrington, R.J., and Spratt, J., 2017. Actively forming Kuroko-type volcanic-hosted massive sulfide (VHMS) mineralization at Iheya North, Okinawa Trough. *Ore Geology Reviews*, 84: 20–41. doi.org/10.1016/j.oregeorev.2016.12.014

Expedition 334: Gary Huftile

Vannucchi, P., Sak, P.B., Morgan, J.P., Ohkushi, K., Ujiie, K., Stroncik, N., Malinverno, A., Arroyo, I., Barckhausen, U., Conin, M.J., Murr Foley, S., Formolo, M.J., Harris, R.N., Heuret, A., **Huftile, G.J.**, and the IODP Expedition 334 Shipboard Scientists, 2013. Rapid pulses of uplift, subsidence, and subduction erosion offshore Central America: Implications for building the rock record of convergent margins. *Geology*, 41(9): 995. doi.org/10.1130/G34355.1

Vannucchi, P., Ujiie, K., Stroncik, N., Malinverno, A., Arroyo, I., Barckhausen, U., Conin, M.J., Murr Foley, S., Formolo, M.J., Harris, R.N., Heuret, A., **Huftile, G.J.**, and the IODP Expedition 334 Scientists, 2011. Costa Rica seismogenesis project (CRISP): Sampling and quantifying input to the seismogenic zone and fluid output. *IODP Preliminary Report*, 334.

Vannucchi, P., Ujiie, K., Stroncik, N., Malinverno, A., Arroyo, I., Barckhausen, U., Conin, M.J., Murr Foley, S., Formolo, M.J., Harris, R.N., Heuret, A., **Huftile, G.J.**, and the IODP Expedition 334 Science Party, 2013. IODP Expedition 334: An investigation of the sedimentary record, fluid flow and state of stress on top of the seismogenic zone of an erosive subduction margin. *Scientific Drilling*, 15: 23–30. doi.org/10.5194/sd-15-23-2013

Vannucchi, P., Ujiie, K., Stroncik, N., Malinverno, A., Arroyo, I., Barckhausen, U., Conin, M.J., Murr Foley, S., Formolo, M.J., Harris, R.N., Heuret, A., **Huftile, G.J.**, and the Expedition 334 Scientists, 2012. *Proceedings of the IODP*, 334: Tokyo (Integrated Ocean Drilling Program Management International, Inc.).

Expedition 335: Graham Baines

Teagle, D.A.H., Ildefonse, B., Blum, P., Guérin, G., Zakharova, N., Abe, N., Abily, B., Adachi, Y., Alt, J.C., Anma, R., **Baines, G.**, and the IODP Expedition 335 Scientists, 2011. Superfast spreading rate crust 4: Drilling gabbro in intact ocean crust formed at a superfast spreading rate. *IODP Preliminary Report*, 335.

Teagle, D.A.H., Ildefonse, B., Blum, P., Guérin, G., Zakharova, N., Abe, N., Abily, B., Adachi, Y., Alt, J.C., Anma, R., **Baines, G.**, and the IODP Expedition 335 Scientists, 2012. IODP Expedition 335: Deep sampling in ODP Hole 1256D. *Scientific Drilling*, 13: 28–34. doi.org/10.5194/sd-13-28-2012

Teagle, D.A.H., Ildefonse, B., Blum, P., Guérin, G., Zakharova, N., Abe, N., Abily, B., Adachi, Y., Alt, J.C., Anma, R., **Baines, G.**, and the Expedition 335 Scientists, 2012. *Proceedings of the IODP*, 335: Tokyo (Integrated Ocean Drilling Program Management International, Inc.).

Expedition 337: Rita Susilawati

Inagaki, F., Hinrichs, K.-U., Kubo, Y., Sanada, Y., Bowden, S., Bowles, M., Glombitza, C., Harrington, G., Heuer, V., Hori, T., Hoshino, T., Ijiri, A., Lever, M.A., Limmer, D., Lin, Y.-S., Liu, C., Morita, S., Morono, Y., Murayama, M., Riedinger, N., Park, Y.-S., Phillips, S., Purkey, M., Reischenbacher, D., Sauvage, J., Snyder, G., **Susilawati, R.**, Tanikawa, W., Trembrath-Reichert, E., Hong, W.-L., and Yamada, Y., 2012. Deep coalbed biosphere off Shimokita: Microbial processes and hydrocarbon system associated with deeply buried coalbed in the ocean. *IODP Preliminary Report*, 337.

Inagaki, F., Hinrichs, K.-U., Kubo, Y., Sanada, Y., Bowden, S., Bowles, M., Glombitza, C., Harrington, G., Heuer, V., Hori, T., Hoshino, T., Ijiri, A., Lever, M.A., Limmer, D., Lin, Y.-S., Liu, C., Morita, S., Morono, Y., Murayama, M., Riedinger, N., Park, Y.-S., Phillips, S., Purkey, M., Reischenbacher, D., Sauvage, J., Snyder, G., **Susilawati, R.**, Tanikawa, W., Trembrath-Reichert, E., Hong, W.-L., and Yamada, Y., 2013. *Proceedings of the IODP*, 337: Tokyo (Integrated Ocean Drilling Program Management International, Inc.).

Expedition 338: Lionel Esteban

Moore, G., Kanagawa, K., Strasser, M., Dugan, B., Maeda, L., Toczko, S., Kido, Y., Thu, M.K., Sanada, Y., **Esteban, L.**, and the IODP Expedition 338 Scientists, 2013. NanTroSEIZE Stage 3: NanTroSEIZE plate boundary deep riser 2. *IODP Preliminary Report*, 338.

Strasser, M., Dugan, B., Kanagawa, K., Moore, G.F., Toczko, S., Maeda, L., Kido, Y., Thu, M.K., Sanada, Y., **Esteban, L.**, and the IODP Expedition 338 Scientists, 2014. *Proceedings from the IODP*, 338: Yokohama (Integrated Ocean Drilling Program).

Expedition 339: Craig Sloss

Dorador, J., Rodríguez-Tovar, F.J., Hernández-Molina, F.J., Stow, D., Alvarez-Zarikian, C., Williams, T., Lofi, J., Acton, G., Bahr, A., Balestra, B., Ducassou, E., Flores, J.-A., Furota, S., Grunert, P., Hodell, D.A., Jimenez-Espejo, F.J., Kom, J.K., Krissek, L.A., Kuroda, J., Li, B., Lorens, L., Miller, M.D., Nanayama, F., Nishida, N., Richter, C., Sanchez Goni, M.F., Sierro Sánchez, F.J., Singh, A.D., **Sloss, C.R.**, and the IODP Expedition 339 Scientists, 2014. Digital image treatment applied to ichnological analysis of marine sediments. *Facies*, 60(1): 39–44. doi.org/10.1007/s10347-013-0383-z

Dorador, J., Rodríguez-Tovar, F.J., Hernández-Molina, F.J., Stow, D., Alvarez-Zarikian, C., Williams, T., Lofi, J., Acton, G., Bahr, A., Balestra, B., Ducassou, E., Flores, J.-A., Furota, S., Grunert, P., Hodell, D.A., Jimenez-Espejo, F.J., Kom, J.K., Krissek, L.A., Kuroda, J., Li, B., Lorens, L., Miller, M.D., Nanayama, F., Nishida, N., Richter, C., Sanchez Goni, M.F., Sierro Sánchez, F.J., Singh, A.D., **Sloss, C.R.**, and IODP Expedition 339 Scientists, 2014. Quantitative estimation of bioturbation based on digital image analysis. *Marine Geology*, 349: 55–60. doi.org/10.1016/j.margeo.2014.01.003

Hernández-Molina F.J., Stow D.A.V., Alvarez-Zarikian C.A., Acton G., Bahr A., Balestra B., Ducassou E., Flood R., Flores J.-A., Furota S., Grunert P., Hodell D., Jimenez-Espejo F., Kim J.K., Krissek L., Kuroda J., Li B., Llave E., Lofi J., Lourens L., Miller M., Nanayama F., Nishida N., Richter C., Roque C., Pereira H., Sanchez Goñi M.F., Sierro F.J., Singh A.D., **Sloss C.**, Takashimizu Y., Tzanova A., Voelker A., Williams T., and Xuan C. 2014. Onset of Mediterranean outflow into the North Atlantic. *Science*, 344: 1244–1250. doi.org/10.1126/science.1251306

Hernández-Molina, F.J., Stow, D., Alvarez-Zarikian, C., Williams, T., Lofi, J., Acton, G., Bahr, A., Balestra, B., Ducassou, E., Flores, J.-A., Furota, S., Grunert, P., Hodell, D.A., Jimenez-Espejo, F.J., Kom, J.K., Krissek, L.A., Kuroda, J., Li, B., Lorens, L., Miller, M.D., Nanayama, F., Nishida, N., Richter, C., Sanchez Goni, M.F., Sierro Sánchez, F.J., Singh, A.D., **Sloss, C.R.**, and Expedition IODP 339 Scientists, 2013. IODP Expedition 339 in the Gulf of Cadiz and off West Iberia: Decoding the environmental significance of the Mediterranean outflow water and its global influence. *Scientific Drilling*, 16: 1–11. doi.org/10.5194/sd-16-1-2013

Hernández-Molina, F.J., Stow, D.A.V., Alvarez Zarikian, C., Williams, T., Lofi, J., Acton, G., Bahr, A., Balestra, B., Ducassou, E., Flores, J.-A., Furota, S., Grunert, P., Hodell, D.A., Jimenez-Espejo, F.J., Kom, J.K., Krissek, L.A., Kuroda, J., Li, B., Lorens, L., Miller, M.D., Nanayama, F., Nishida, N., Richter, C., Sanchez Goni, M.F., Sierro Sánchez, F.J., Singh, A.D., **Sloss, C.R.**, and the IODP Expedition 339 Scientists, 2012. Mediterranean outflow: Environmental significance of the Mediterranean Outflow Water and its global implications. *IODP Preliminary Report*, 339.

Hodell, D., Lourens, L., Crowhurst, S., Konijnendijk, T., Tjallingii, R., Jiménez-Espejo, F., Skinner, L., Tzedakis, P.C., Abrantes, F., Acton, G.D., Alvarez Zarikian, C.A., Bahr, A., Balestra, B., Barranco, E.L., Carrara, G., Ducassou, E., Flood, R.D., Flores, J.-A., Furota, S., Grimalt, J., Grunert, P., Hernández-Molina, J., Kim, J.K., Krissek, L.A., Kuroda, J., Li, B., Lofi, J., Margari, V., Martrat, B., Miller, M.D., Nanayama, F., Nishida, N., Richter, C., Rodrigues, T., Rodríguez-Tovar, F.J., Roque, A.C.F., Sanchez Goni, M.F., Sierro Sánchez, F.J., Singh, A.D., **Sloss, C.R.**, Stow, D.A.V., Takashimizu, Y., Tzanova, A., Voelker, A., Xuan, C., and Williams, T., 2015. A reference time scale for Site U1385 (Shackleton Site) on the SW Iberian Margin. *Global and Planetary Change*, 133: 49–64. doi.org/10.1016/j.gloplacha.2015.07.002

Hodell, D.A., Lourens, L., Stow, D.A.V., Hernández-Molina, J., Alvarez Zarikian, C.A., **Sloss, C.R.**, and the Shackleton Site Project Members: The 'Shackleton Site' (IODP Site U1385) on the Iberian Margin. *Scientific Drilling*, 16: 13–19. doi.org/10.5194/sd-16-13-2013

Stow, D.A.V., Hernández-Molina, F.J., Alvarez Zarikian, C.A., Williams, T., Lofi, J., Acton, G., Bahr, A., Balestra, B., Ducassou, E., Flores, J.-A., Furota, S., Grunert, P., Hodell, D.A., Jimenez-Espejo, F.J., Kom, J.K., Krissek, L.A., Kuroda, J., Li, B., Lorens, L., Miller, M.D., Nanayama, F., Nishida, N., Richter, C., Sanchez Goni, M.F., Sierro Sánchez, F.J., Singh, A.D., **Sloss, C.R.**, and the IODP Expedition 339 Scientists, 2013. *Proceedings of the IODP*, 339: Tokyo (Integrated Ocean Drilling Program Management International, Inc.).

Expedition 340: Martin Jutzeler

Cassidy M., Watt, S.F.L., Talling, P.J., Palmer, M.R., Edmonds, M., Jutzeler, M., Wall-Palmer, D., Manga, M., Coussens, M., Gernon, T., Taylor, R.N., Michalik, A., Inglis, E., Breitkreuz, C., Le Friant, A., Ishizuka, O., Boudon, G., McCanta, M.C., Adachi, T., Hornbach, M.J., Colas, S.L., Endo, D., Fujinawa, A., Kataoka, K.S., Maeno, F., Tamura, Y. and Wang, F., 2015. Magmatism following unloading by volcanic edifice collapse is mafic, deep and rapid. *Geophysical Research Letters*, 42: 4778–4785. doi.org/10.1002/2015GL064519

Coussens, M.F., Wall-Palmer, D., Talling, P.J., Watt, S.F.L., Hatter, S.J., Cassidy, M., Clare, M., **Jutzeler, M.**, Hatfield, R., McCanta, M., Kataoka, K.S., Endo, D., Palmer, M.R., Stinton, A., Fujinawa, A., Boudon, G., Le Friant, A., Ishizuka, O., Gernon, T., Adachi, T., Aljahdali, M., Breitkreuz, C., Frass, A.J., Hornbach, M.J, Lebas, E., Lafuerza, S., Maeno, F., Manga, M., Martinez-Colon, M., McManus, J., Morgan, S., Saito, T., Slagle, A., Subramanyam, K.S.V., Tamura, Y., Trofimovs, J., Villemant, B., Wang, F., and the Expedition 340 scientists, 2015. Synthesis: Stratigraphy and age control for IODP Sites U1394, U1395 and U1396 offshore Montserrat in the Lesser Antilles. *Proceedings of the IODP*, 340

Jutzeler, M., White, J.D.L., Talling, P.J., McCanta, M., Morgan, S., Le Friant, A., and Ishizuka, O., 2014. Coring disturbances in IODP piston cores with implications for offshore record of volcanic events and the Missoula megafloods. *G3: Geochemistry, Geophysics, Geosystems*, 15(9): 3572–3590. doi.org/10.1002/2014GC005447

Jutzeler M., Talling, P.J., White, J.D.L., and the Expedition 340 Scientists, 2016. Data report: Coring disturbances in IODP 340, a detailed list of intervals with fall-in and flow-in. *Proceedings of the IODP*, 340. doi.org/10.2204/iodp. proc.340.206.2016

Le Friant, A., Ishizuka, O., Boudon, G., Palmer, M.R., Talling, P.J., Villemant, B., Adachi, T., Aljahdali, M., Breitkreuz, C., Brunet, M., Caron, B., Coussens, M., Deplus, C., Endo, D., Feuillet, N., Fraas, A.J., Fujinawa, A., Hart, M.B., Hatfield, R.B., Hornbach, M., **Jutzeler, M.**, Kataoka, K.S., Komorowski, J.-C., Lebas, E., Lafuerza, S., Maeno, F., Manga, M., Martínez-Colon, M., McCanta, M., Morgan, S., Saito, T., Slagle, A., Sparks, S., Stinton, A., Stroncik, N., Subramanyam, K.S.V., tamura, Y., Trofimovs, J., Voight, B., Wall-Palmer, D., Wang, F., and Watt, S.F.L., 2015. Submarine record of volcanic island construction and collapse in the Lesser Antilles arc: first scientific drilling of submarine volcanic island landslides by IODP Expedition 340. *G3: Geochemistry, Geophysics, Geosystems*, 16(2): 420–442. doi.org/10.1002/2014GC005652

Le Friant, A., Ishizuka, O., Stroncik, N.A., Slagle, A.L., Morgan, S., Adachi, T., Aljahdali, M., Boudon, G., Breikreuz, C., Endo, D., Fraass, A.J., Fujinawa, A., Hatfield, R.G., Hornbach, M.J., **Jutzeler, M.**, and the IODP Expedition 340 Scientists, 2012. Lesser Antilles volcanism and landslides: implications for hazard assessment and long-term magmatic evolution of the arc. *IODP Preliminary Report*, 340.

Le Friant, A., Ishizuka, O., Stroncik, N.A., Slagle, A.L., Morgan, S., Adachi, T., Aljahdali, M., Boudon, G., Breikreuz, C., Endo, D., Fraass, A.J., Fujinawa, A., Hatfield, R.G., Hornbach, M.J., **Jutzeler, M.**, and the Expedition 340 Scientists, 2013. *Proceedings of the IODP*, 340: Tokyo (Integrated Ocean Drilling Program Management International, Inc.).

Manga, M., Hornbach, M.J., Le Friant, A., Ishizuka, O., Stroncik, N., Adachi, T., Aljahdali, M., Boudon, G., Breitkreuz, C., Fraass, A., Fujinawa, A., Hatfield, R., **Jutzeler, M.**, Kataoka, K., Lafuerza, S., Maeno, F., Martinez-Colon, M., McCanta, M., Morgan, S., Palmer, M.R., Saito, T., Slagle, A., Stinton, A.J., Subramanyam, K.S.V., Tamura, Y., Talling, P.J., Villemant, B., Wall-Palmer, D., and Wang, F., 2012. Heat flow in the Lesser Antilles island arc and adjacent back arc Grenada Basin. *G3: Geochemistry, Geophysics, Geosystems*, 13. doi.org/10.1029/2012gc004260

Wall-Palmer, D., Coussens, M., Talling, P.J., **Jutzeler, M.**, Cassidy, M., Marchant, I., Palmer, M.R., Watt, S.F.L., Smart, C.W., Fisher, J.K., Hart, M.B., Fraass, A., Trofimovs, J., Le Friant, A., Ishizuka, O., Adachi, T., Aljahdali, M., Boudon, G., Breitkreuz, C., Endo, D., Fujinawa, A., Hatfield, R., Hornbach, M.J., Kataoka, K., Lafuerza, S., Maeno, F., Manga, M., Martinez-Colon, M., McCanta, M., Morgan, S., Saito, T., Slagle, A.L., Stinton, A.J., Subramanyam, K.S.V., Tamura, Y., Villemant, B., and Wang, F., 2014. Late Pleistocene stratigraphy of IODP Site U1396 and compiled chronology offshore of south and south west Montserrat, Lesser Antilles. *G3: Geochemistry, Geophysics, Geosystems*, 15(7): 3000–3020. doi.org/10.1002/2014GC005402

Expedition 341: Christopher Moy, Maureen Walczak (née Davies), Carol Larson

Gulick, S.P., J.M. Jaeger, A.C. Mix, H. Asahi, H. Bahlburg, C. Belanger, G.B.B. Berbel, L. Childress, E. Cowan, M.H. **Davies-Walczak, L.** Drab, F. Dottori, M. Forwick, A. Fukumura, S. Ge, S. Gupta, A. Kioka, S. Konno, L. LeVay, C. März, K. Matsuzaki, E. McClymont, **C. Moy**, J. Müller, A. Nakamura, T. Ojima, K. Ridgeway, O. Romero, A. Slagle, J. Stoner, G. St-Onge, I. Suto, L. Worthington, I. Bailey, E. Enkelmann, and R. Reece, 2015. Nonlinear feedback between tectonic uplift and glacial erosion constrained by Gulf of Alaska sedimentary record. *Proceedings of the National Academy of Sciences,* 112(49): 15042–15047. doi.org/10.1073/pnas.1512549112

Jaeger, J.M., Gulick, S.S., LeVay, L.J., Slagle, A.L., Drab, L., Asashi, H., Bahlburg, H., Belanger, C.L., Berbel, G.B.B., Childress, L.B., Cowan, E.A., Forwick, M., Fukumura, A., Ge, S., Gupta, S.M., Kioka, A., Konno, S., Marz, C.E., Matsuzaki, K.M., McClymont, E.L., Mix, A.C., **Moy, C.M.**, Muller, J., Nakamura, A., Ojima, T., Ridgway, K.D., Rodrigues Ribeiro, F., Romero, O.E., Stoner, J.S., St-Onge, G., Suto, I., **Walczak, M.H.**, Worthington, L.L., and **Larson, C.**, 2014. Southern Alaska margin: interactions of tectonics, climate, and sedimentation. *IODP Preliminary Report*, 341.

Jaeger, J.M., Gulick, S.S., LeVay, L.J., Slagle, A.L., Drab, L., Asashi, H., Bahlburg, H., Belanger, C.L., Berbel, G.B.B., Childress, L.B., Cowan, E.A., Forwick, M., Fukumura, A., Ge, S., Gupta, S.M., Kioka, A., Konno, S., Marz, C.E., Matsuzaki, K.M., McClymont, E.L., Mix, A.C., Moy, C.M., Muller, J., Nakamura, A., Ojima, T., Ridgway, K.D., Rodrigues Ribeiro, F., Romero, O.E., Stoner, J.S., St-Onge, G., Suto, I., **Walczak, M.H.**, Worthington, L.L., **Larson, C.**, 2014. *Proceedings of the IODP*, 341: College Station, TX (Integrated Ocean Drilling Program).

Montelli, A., Gulick S.P., Worthington L.L., Mix A., **Davies-Walczak M.H.**, Zellers S., Jaeger J. (2017), Late Quaternary glacial dynamics and sedimentation variability in Bering Trough, Gulf of Alaska. *Geology*, 45(1). doi.org/10.1130/G38836.1

Walczak, M.H., Mix, A.C., Willse, T., Slagle, A., Stoner, J.S., Jaeger, J., Gulick, S., LeVay, L., Kioka, A., and the IODP Expedition 341 Scientific Party, 2015. Correction of non-intrusive drill core physical properties data for variability in recovered sediment volume. *Geophysical Journal International* 202(2): 1317–1323. doi.org/10.1093/gji/ggv204

Expedition 342: Brad Opdyke, Chris Hollis

Flemings, P.B., Polito, P.J., Pettigrew, T.L., Iturrino, G.J., Meissner, E., Aduddell, R., Brooks, D.L., Hetmaniak, C., Huey, D., Germaine, J.T., Norris, R.D., Wilson, P.A., Blum, P., Fehr, A., Agnini, C., Nornemann, A., Boulila, S., Bown, P.R., Cournede, C., Friedrich, O., Ghosh, A.K., **Hollis, C.J.**, Hull, P.M., Jo, K., Junium, C.K., Kaneko, M., Liebrand, D., Lippert, P.C., Liu, Z., Matsui, H., Moriya, K., Nishi, H., **Opdyke, B.N.**, and the IODP Expedition 342 Scientists, 2013. The Motion Decoupled Delivery System: A new deployment system for downhole tools is tested at the New Jersey Margin. *Scientific Drilling*, 15: 51–56. doi.org/10.5194/sd-15-51-2013

Friedrich, O., Norris, R.D., Wilson, P.A., and **Opdyke, B.N.**, 2015. Newfoundland Neogene sediment drifts: transition from the Paleogene greenhouse to the modern icehouse. *Scientific Drilling*, 19, 39–42. doi.org/10.5194/sd-19-39-2015

Hollis, C.J., 2017. Data report: Siliceous microfossil abundance in IODP Expedition 342 sediments. *Proceedings of the IODP*, 342. doi.org/10.2204/iodp.proc.342.201.2017

Norris, R.D., Wilson, P.A, Blum, P., Fehr, A., Agnini, C., Nornemann, A., Boulila, S., Bown, P.R., Cournede, C., Friedrich, O., Ghosh, A.K., **Hollis, C.J.**, Hull, P.M., Jo, K., Junium, C.K., Kaneko, M., Liebrand, D., Lippert, P.C., Liu, Z., Matsui, H., Moriya, K., Nishi, H., **Opdyke, B.N.**, and the IODP Expedition 342 Scientists, 2014. Proc*eedings of the IODP*, 342: College Station, TX (Integrated Ocean Drilling Program).

Norris, R.D., Wilson, P.A., Blum, P., Fehr, A., Agnini, C., Nornemann, A., Boulila, S., Bown, P.R., Cournede, C., Friedrich, O., Ghosh, A.K., **Hollis, C.J.**, Hull, P.M., Jo, K., Junium, C.K., Kaneko, M., Liebrand, D., Lippert, P.C., Liu, Z., Matsui, H., Moriya, K., Nishi, H., **Opdyke, B.N.**, and the IODP Expedition 342 Scientists, 2012. Paleogene Newfoundland sediment drifts. *IODP Preliminary Report*, 342.

Expedition 343: Virginia Toy

Bose S., Saha P., Mori J.J., Rowe C., Ujiie K., Chester F.M., Conin M., Regalla C., Kameda J., **Toy V.**, Kirkpatrick J., Remitti F., Moore J.C., Wolfson-Schwehr M., Nakamura Y., and Gupta A., 2015. Deformation structures in the frontal prism near the Japan Trench: Insights from sandbox models. *Journal of Geodynamics*, 89: 29–38. doi.org/10.1016/j.jog.2015.06.002

Chester, F.M., Mori, J., Eguchi, N., Toczko, S., Kido, Y., Saito, S., Sanada, Y., Anderson, L., Behremann, J.H., Bose, S., Conin, M., Cook, B., Fulton, P., Hirose, T., Ikari, M., Ishikawa, T., Jeppson, Kameda, J., Kirkpatrick, J., Lin, W., Mishima, T., Moore, J.C., Nakamura, Y., Regalla, C., Remitti, F., Rowe, C., Sample, J., Sun, T., Takai, K., **Toy, V.**, and the Expedition 343/343T Scientists, 2013. *Proceedings of the IODP*, 343/343T: Tokyo (Integrated Ocean Drilling Program Management International, Inc.).

Chester, F.M., Mori, J.J., Toczko, S., Eguchi, N., Kido, Y., Saito, S., Sanada, Y., Anderson, L., Behremann, J.H., Bose, S., Conin, M., Cook, B., Fulton, P., Hirose, T., Ikari, M., Ishikawa, T., Jeppson, Kameda, J., Kirkpatrick, J., Lin, W., Mishima, T., Moore, J.C., Nakamura, Y., Regalla, C., Remitti, F., Rowe, C., Sample, J., Sun, T., Takai, K., **Toy, V.**, Ujiie, K., Wolfson, M., and Yang, T., 2012. Japan Trench Fast Drilling Project (JFAST). *IODP Preliminary Report*, 343/343T.

Chester, F.M., Rowe, C., Ujiie, K., Kirkpatrick, J., Regalla, C., Remitti, F., Moore, J.C., **Toy, V.**, Wolfson-Schwehr, M., Bose, S., Kameda, J., Mori, J.J., Brodsky, E.E., Eguchi, N., Toczko, S., and the Expedition 343 and 343T Scientists, 2013. Structure and composition of the plate-boundary slip zone for the 2011 Tohoku-oki Earthquake. *Science*, 342(6163): 1208–1211. doi.org/10.1126/science.1243719

Kirkpatrick, J.D., Rowe, C.D., Ujiie, K., Moore, J.C., Regalla, C., Remitti, F., **Toy, V.**, Wolfson-Schwehr, M., Kameda, J., Bose, S., and Chester, F.M., 2015. Structure and lithology of the Japan Trench subduction plate boundary fault. *Tectonics*, 34: 53–69. doi.org/10.1002/2014TC003695

Lin W., Conin M., Moore J.C., Chester F M., Nakamura Y., Mori J.J., Anderson L., Brodsky E.E., Eguchi N., Toczko, S., Kido, Y., Saito, S., Sanada, Y., Behremann, J.H., Bose, S., Cook, B., Fulton, P., Hirose, T., Ikari, M., Ishikawa, T., Jeppson, Kameda, J., Kirkpatrick, J., Mishima, T., Regalla, C., Remitti, F., Rowe, C., Sample, J., Sun, T., Takai, K., **Toy, V.**, Ujiie, K., Wolfson, M., and Yang, T., 2013. Stress State in the Largest Displacement Area of the 2011 Tohoku-Oki Earthquake. *Science*, 339: 687–690. doi.org/10.1126/science.1229379

Expedition 344: Alan Baxter

Harris, R.N., Sakaguchi, A., Petronotis, K., Malinverno, A., **Baxter, A.T.**, and the IODP Expedition 344 Scientists, 2013. Costa Rica Seismogenesis Project, Program A Stage 2 (CRISP-A2): Sampling and quantifying lithologic inputs and fluid inputs and outputs of the seismogenic zone. *IODP Preliminary Report*, 344.

Harris, R.N., Sakaguchi, A., Petronotis, K., Malinverno, A., **Baxter, A.T.**, and the Expedition 344 Scientists, 2013. *Proceedings of the IODP*, 344: College Station, TX (Integrated Ocean Drilling Program).

Schindlbeck, J.C., Kutterolf, S., Freundt, A., Straub, S.M., Wang, K.-L., Jegen, M., Hemming, S.R., **Baxter, A.T.**, and Sandoval, M.I., 2015. The Miocene Galápagos ash layer record of Integrated Ocean Drilling Program Legs 334 and 344: Ocean-island explosive volcanism during plume-ridge interaction. *Geology*, 43(7): 599–602. doi.org/10.1130/G36645.1

Expedition 345: Trevor Falloon

Gillis, K.M., Snow, J.E., Klaus, A., Guerin, G., Abe, N., Akizawa, N., Ceuleneer, G., Cheadle, M.J., Adrião, A.B., Faak, K., **Falloon, T.J.**, and the IODP Expedition 345 Scientists, 2013. Hess Deep plutonic crust: Exploring the plutonic crust at a fast-spreading ridge: new drilling at Hess Deep. *IODP Preliminary Report*, 345.

Gillis, K.M., Snow, J.E., Klaus, A., Guerin, G., Abe, N., Akizawa, N., Ceuleneer, G., Cheadle, M.J., Adrião, A.B., Faak, K., **Falloon, T.J.**, and the Expedition 345 Scientists, 2014. *Proceedings of the IODP*, 345: College Station, TX (Integrated Ocean Drilling Program).

Gillis, K.M., Snow, J.E., Klaus, A., Abe, N., Adrião, Á.B., Akizawa, N., Ceuleneer, G., Cheadle, M.J., Faak, K., **Falloon, T.J.**, Friedman, S.A., Godard, M., Guerin, G., Harigane, Y., Horst, A.J., Hoshide, T., Ildefonse, B., Jean, M.M., John, B.E., Koepke, J., Machi, S., Maeda, J., Marks, N.E., McCaig, A.M., Meyer, R., Morris, A., Nozaka, T., Python, M., Saha, A., and Wintsch, R.P., 2013. Primitive layered gabbros from fast-spreading lower oceanic crust. *Nature*, 505(7482): 204–207. doi.org/10.1038/nature12778

Expedition 346: Stephen Gallagher

Tada, R., Murray, R.W., Alvarez Zarikian, C.A., Lofi, J., Anderson, W.T., Bassetti, M.-A., Brace, B.J., Clemens, S.C., Dickens, G.R., Dunlea, A.G., **Gallagher, S.J.**, and the IODP Expedition 346 Scientists, 2013. Asian Monsoon: Onset and evolution of millennial-scale variability of Asian monsoon and its possible relation with Himalaya and Tibetan Plateau uplift. *IODP Preliminary Report*, 346.

Tada, R., Murray, R.W., Alvarez Zarikian, C.A., Lofi, J., Anderson, W.T., Bassetti, M.-A., Brace, B.J., Clemens, S.C., Dickens, G.R., Dunlea, A.G., **Gallagher, S.J.**, and the Expedition 346 Scientists, 2015. *Proceedings of the IODP*, 346: College Station, TX (Integrated Ocean Drilling Program).

Expedition 348: Matthew Josh

Tobin, H., Hirose, T., Saffer, D., Toczko, S., Maeda, L., Kubo, Y., Sanada, Y., Kido, Y., Hamada, Y., Boston, B., Broderick, A., Brown, K., Crespo-Blanc, A., Even, E., Fuchida, S., Fukuchi, R., Hammerschmidt, S., Henry, P., **Josh, M.**, and the IODP Expedition 348 Scientists and Scientific Participants, 2014. NanTroSEIZE Stage 3: NanTroSEIZE plate boundary deep riser 3. *IODP Preliminary Report*, 348.

Tobin, H., Hirose, T., Saffer, D., Toczko, S., Maeda, L., Kubo, Y., Kido, Y., Hamada, Y., Boston, B., Broderick, A., Brown, K., Crespo-Blanc, A., Even, E., Fuchida, S., Fukuchi, R., Hammerschmidt, S., Henry, P., **Josh, M.**, and the Expedition 348 Scientists, 2015. *Proceedings of the IODP*, 348: College Station, TX (Integrated Ocean Drilling Program).

Partners in the first phase of IODP: 2008–2013

Australian Institutions

Australian Institute of Marine Science
The Australian National University
Australian Nuclear Science and Technology Organisation
CSIRO Earth Science and Resource Engineering
Curtin University of Technology
Geoscience Australia
James Cook University
Macquarie University
Monash University
Queensland University of Technology
University of Adelaide
University of Melbourne
University of Newcastle
University of New England
University of Queensland
University of Sydney
University of Tasmania
University of Technology Sydney
University of Western Australia
University of Wollongong
MARGO (Marine Geoscience Office)

New Zealand Institutions

GNS Science
University of Auckland
University of Otago
Victoria University of Wellington